# BIOBLEACHING

# BIOBLEACHING

**Edited by:**

**Shiv Sanjeevi and Dr Prerna Pandey**

Delve
PUBLISHING

www.delvepublishing.com

**Biobleaching**

*Shiv Sanjeevi and Dr Prerna Pandey*

**Delve Publishing**

2010 Winston Park Drive,

2nd Floor

Oakville, ON L6H 5R7

Canada

www.delvepublishing.com

Tel: 001-289-291-7705
001-905-616-2116
Fax: 001-289-291-7601

Email: orders@arclereducation.com



ISBN: 978-1-77361-189-1 (Hardcover)

# DECLARATION

# ABOUT THE EDITORS

**Shiv Sanjeevi Sripathi** completed his BSc in Biotechnology from Mumbai University and secured second rank in Mumbai University and first rank in his college: Kishinchand chellaram College. Shiv secured first rank in Masters his college KET's V.G.Vaze College and has worked on a stem cell project at the Specialized Centre for Cell Based Therapy (SCCT), KEM Hospital at Mumbai, following which he was a Junior Research Fellow at Microbiology & Cell Biology Department, Indian Institute of Science, Bangalore from August 2009 to November 2011. The field of work involved cloning of cell wall genes and transcription factors in *E. coli & M. smegmatis*. This was followed by a stint as a Research and Development Executive at Transasia Biomedicals, Mumbai where he worked on patent writing and developing immunoassays from 2011 till 2013. In 2013, he entered the teaching field where he has been tutoring biology and chemistry to the students of 11th and 12th grades along with English grammar and communication skills to people of all age groups.

**Dr Prerna Pandey,** a biotechnologist with several years of wet lab research experience, worked at International Center for Genetic Engineering and Biotechnology, New Delhi. She has also worked at Transasia Biomedicals and Advance Enzyme Technologies as a scientist. Prerna has published papers in peer-reviewed journals, and has submitted a number of annotated Geminiviral genome sequences in the GenBank, including two novel ones. Now, she works as a freelance scientific editor and writer.

# TABLE OF CONTENTS

# LIST OF CONTRIBUTORS

**Shiv Sanjeevi Sripathi**

Shiv Sanjeevi Sripathi completed his BSc in Biotechnology from Mumbai University and secured second rank in Mumbai University and first rank in his college: Kishinchand chellaram College. Shiv secured first rank in his college KET's V.G.Vaze College and has worked on a stem cell project at the Specialized Centre for Cell Based Therapy (SCCT), KEM Hospital at Mumbai, following which he was a Junior Research Fellow at Microbiology & Cell Biology Department, Indian Institute of Science, Bangalore from August 2009 to November 2011. The field of work involved cloning of cell wall genes and transcription factors in E. coli & M. smegmatis. This was followed by a stint as a Research and Development Executive at Transasia Biomedicals, Mumbai where he worked on patent writing and developing immunoassays from 2011 till 2013. In 2013, he entered the teaching field where he has been tutoring biology and chemistry to the students of 11th and 12th grades along with English grammar and communication skills to people of all age groups.

**Prerna Pandey**

Dr Prerna Pandey, a biotechnologist with several years of wet lab research experience, worked at International Center for Genetic Engineering and Biotechnology, New Delhi. She has also worked at Transasia Biomedicals and Advance Enzyme Technologies as a scientist. Prerna has published papers in peer-reviewed journals, and has submitted a number of annotated Geminiviral genome sequences in the GenBank, including two novel ones. Now, she works as a freelance scientific editor and writer.

**Paolo Zucca**

Dipartimento di Scienze Biomediche, Università di Cagliari, Cittadella Universitaria, 09042, Monserrato, CA, Italy

Consorzio UNO (University of Oristano Consortium), Oristano, Italy

**Antonio Rescigno**

Dipartimento di Scienze Biomediche, Università di Cagliari, Cittadella Universitaria, 09042, Monserrato, CA, Italy

**Manuela Pintus**
Dipartimento di Scienze Biomediche, Università di Cagliari, Cittadella Universitaria, 09042, Monserrato, CA, Italy

**Andrea C Rinaldi**
Dipartimento di Scienze Biomediche, Università di Cagliari, Cittadella Universitaria, 09042, Monserrato, CA, Italy

**Enrico Sanjust**
Dipartimento di Scienze Biomediche, Università di Cagliari, Cittadella Universitaria, 09042, Monserrato, CA, Italy

**Cun-Duo Tang**
School of Biotechnology and Key Laboratory of Industrial Biotechnology, Ministry of Education, Jiangnan University, Wuxi, Jiangsu, China

**Jian-Fang Li**
School of Food Science and Technology, Jiangnan University, Wuxi, Jiangsu, China

**Xi-Huan Wei**
School of Food Science and Technology, Jiangnan University, Wuxi, Jiangsu, China

**Rou Min**
School of Pharmaceutical Science, Jiangnan University, Wuxi, Jiangsu, China

**Shu-Juan Gao**
School of Pharmaceutical Science, Jiangnan University, Wuxi, Jiangsu, China

**Jun-Qing Wang**
School of Biotechnology and Key Laboratory of Industrial Biotechnology, Ministry of Education, Jiangnan University, Wuxi, Jiangsu, China

**Xin Yin**
School of Biotechnology and Key Laboratory of Industrial Biotechnology, Ministry of Education, Jiangnan University, Wuxi, Jiangsu, China

**Min-Chen Wu**
Wuxi Medical School, Jiangnan University, Wuxi, Jiangsu, China

**Antar Puneet Virk**
Department of Biotechnology, Panjab University, Chandigarh, India

**Minakshi Puri**
Department of Microbiology, Panjab University, Chandigarh, India

**Vijaya Gupta**
Department of Microbiology, Panjab University, Chandigarh, India

**Neena Capalash**
Department of Biotechnology, Panjab University, Chandigarh, India

**Prince Sharma**
Department of Microbiology, Panjab University, Chandigarh, India

**Marie Couturier**
INRA, UMR1163, Laboratoire de Biotechnologie des Champignons Filamenteux, Polytech Marseille, Marseille, France
Aix Marseille Universite´, Polytech Marseille, Marseille, France

**Julia Fe´liu**
INRA, UMR1163, Laboratoire de Biotechnologie des Champignons Filamenteux, Polytech Marseille, Marseille, France
Aix Marseille Universite´, Polytech Marseille, Marseille, France

**Sophie Bozonnet**
Universite´ de Toulouse; INSA, UPS, INP; LISBP, Toulouse, France
INRA, UMR792, Inge´nierie des Syste`mes Biologiques et des Proce´de´s, Toulouse, France
CNRS, UMR5504, Toulouse, France

**Alain Roussel**
Architecture et Fonction des Macromole´cules Biologiques, UMR7257, CNRS, Aix Marseille Universite´, Marseille, France

**Jean-Guy Berrin**
INRA, UMR1163, Laboratoire de Biotechnologie des Champignons Filamenteux, Polytech Marseille, Marseille, France
Aix Marseille Universite´, Polytech Marseille, Marseille, France

**Wangning Yu**
College of Biological Sciences and Technology, Beijing Forestry University, Beijing, PR China

**Weina Liu**
College of Biological Sciences and Technology, Beijing Forestry University, Beijing, PR China

**Huoqing Huang**
Feed Research Institute Chinese Academy of Agricultural Sciences, Beijing, PR China

**Fei Zheng**
College of Biological Sciences and Technology, Beijing Forestry University, Beijing, PR China

**Xiaoyu Wang**
College of Biological Sciences and Technology, Beijing Forestry University, Beijing, PR China

**Yuying Wu**
College of Materials Science and Technology, Beijing Forestry University, Beijing, PR China

**Kangjia Li**
College of Biological Sciences and Technology, Beijing Forestry University, Beijing, PR China

**Xiangming Xie**
College of Biological Sciences and Technology, Beijing Forestry University, Beijing, PR China

**Yi Jin**
College of Biological Sciences and Technology, Beijing Forestry University, Beijing, PR China

**Saurabh Sudha Dhiman**
Department of Chemical Engineering, Konkuk University, Gwangjin–Gu, Seoul, Republic of Korea
Institute of SK–KU Biomaterials, Konkuk University, Gwangjin–Gu, Seoul, Republic of Korea

**Gaurav Garg**
Department of Biotechnology, Kurukshetra University, Kurukshetra, Haryana, India
Department of Biotechnology, Maharishi Markandeshwar University, Mullana-Ambala, Haryana, India

**Jitender Sharma**
Department of Biotechnology, Kurukshetra University, Kurukshetra, Haryana, India

**Vipin C. Kalia**
Microbial Biotechnology and Genomics, CSIR-Institute of Genomics and Integrative Biology, Delhi University Campus, Delhi, India

**Yun Chan Kang**
Department of Chemical Engineering, Konkuk University, Gwangjin–Gu, Seoul, Republic of Korea

**JungKul Lee**
Department of Chemical Engineering, Konkuk University, Gwangjin–Gu, Seoul, Republic of Korea
Institute of SK–KU Biomaterials, Konkuk University, Gwangjin–Gu, Seoul, Republic of Korea

**Digvijay Verma**
Department of Microbiology, University of Delhi South Campus, New Delhi, India

**Yutaka Kawarabayasi**
Laboratory for Functional Genomics of Extremophiles, Faculty of Agriculture, Kyushu University, Fukuoka, Japan
Institute for Hearth Research, National Institute of Advanced Industrial Science and Technology (AIST), Hyogo, Japan

**Kentaro Miyazaki**
Bioproduction Research Institute, National Institute of Advanced Industrial Science and Technology (AIST), Hokkaido, Japan
Department of Medical Genome Sciences, Graduate School of Frontier Sciences, The University of Tokyo, Hokkaido, Japan

**Tulasi Satyanarayana**
Department of Microbiology, University of Delhi South Campus, New Delhi, India

# PREFACE

The use of bacteria or enzymes or biological agents in the removal of color is termed as biobleaching. The employment of various chemicals in manufacturing processes of paper, clothes, coir, jute pose several harsh effects on the environment. In today's world of increasing awareness on environmental safety, the use of nature's selected agents can serve as a safe alternative promising a better world for the future generations.

Research in the field of biobleaching has examined the use of several enzymes produced by bacteria or fungi or biotechnology or microbes themselves or even a cocktail of nature's harvest as microbes and their enzymes to remove color.

This book serves to address various aspects of this field including the history, research conducted, various agents used for biobleaching and also present a merit-demerit analysis and scope of the subject. The authors seek to do their part to environmental safety as this book offers a comprehensive account of the use of nature's agents in decolorization, thereby, leaving a world with color for generations ahead.

Editors

CHAPTER 1

# AN INTRODUCTION TO BIOBLEACHING

**Shiv Sanjeevi Sripathi[1] and Prerna Pandey[2]**

[1]Shiv Sanjeevi Sripathi completed his BSc in Biotechnology from Mumbai University and secured second rank in Mumbai University and first rank in his college: Kishinchand chellaram College. Shiv secured first rank in his college KET's V.G.Vaze College and has worked on a stem cell project at the Specialized Centre for Cell Based Therapy (SCCT), KEM Hospital at Mumbai, following which he was a Junior Research Fellow at Microbiology & Cell Biology Department, Indian Institute of Science, Bangalore from August 2009 to November 2011. The field of work involved cloning of cell wall genes and transcription factors in E. coli & M. smegmatis. This was followed by a stint as a Research and Development Executive at Transasia Biomedicals, Mumbai where he worked on patent writing and developing immunoassays from 2011 till 2013. In 2013, he entered the teaching field where he has been tutoring biology and chemistry to the students of 11th and 12th grades along with English grammar and communication skills to people of all age groups.

[2]Dr Prerna Pandey, a biotechnologist with several years of wet lab research experience, worked at International Center for Genetic Engineering and Biotechnology, New Delhi. Her field of research in PhD included isolation and molecular characterization of geminiviruses, genome sequencing, gene annotation, and gene silencing using the RNA interference technology. She has also worked at Transasia Biomedicals and Advance Enzyme Technologies as a scientist. Prerna has published papers in peer-reviewed journals, and has submitted a number of annotated Geminiviral genome sequences in the GenBank, including two novel ones. She has also completed her editing and proof-reading courses from Society for the promotion of Editors and Proof-readers, UK and now works as a freelance scientific editor and writer. When Prerna is not busy with her assignments, she loves to explore historical places.

## WHERE DO WE USE BLEACH?

The application of biotechnology has been extended to the field of using alternative sources of a particular purpose. With much emphasis being placed on environmental issues, the use of tools of biotechnology offers a viable alternative to minimize the damage to the ecology. Several industries such as textiles and food processing units use dyes that require a bleaching step, along with paper industry that uses bleaching steps.

The wood is composed of cellulose followed by xylan and lignin. The lignin component is a phenol based polymer that is an integral part of most wood that finds use in the paper industry. The use of a bleaching step is important to remove the color impacted by the lignin in the paper and pulp industry (Viikari et al. 1986; Koponen 1991).

The pulp and paper industry along with other industries are working on several aspects of using technologies to minimize the use of chemicals conventionally used in bleaching. Research traces the use of enzymes for removing lignin, called "delignification" was discussed by Viikari et al. (1994). The use of the enzyme xylanase was found to reduce the amount of chlorine used in bleaching.

## WHY BIOBLEACHING?

Color removal can use three methods: biological, chemical and physical. Each method offers its own set of merits and challenges howver the use of biological methods offers several advantages the most important being their biodegradable nature. Several biomass such as rice husk (De et al, 2009). fungi (especially white rot fungi), bacteria, algae have shown promising results in the bleaching of dyes.

Bioremediation and bleaching employing the use of bacteria, yeasts, algae and fungi show much potential in the accumulation and degradation of many dyes in order to achieve biobleaching (Archna et al, 2012 and Kiro et al, 2016).

It was reported by a study that chlorinated organic compounds can be of an immense variety in the effluent of the bleaching industry (Sharma et al. 1997). Several groups include chlorinated resin acids, chlorinated phenolics and dioxin that are associated with notable markers of pollution such as the oxygen demand, color and toxic effects. Ministry of environment and Forest, India listed the paper pulp industry as a leading cause of pollution. This spawned the development in several alternative strategies such as ECF (elemental chlorine free), TCF (total chlorine free) and TEF (totally effluent free) for bleaching.

Several chemicals have been employed for the bleaching in industry that shall be discussed below. However, the use of biological agents such as microbes or enzymes offer several advantages over other methods employed:

- They are products of nature, hence are biodegradable.
- The majority of enzymes are extracellular and fermentation can be used to produce them.

(reviewed by Ghosh et al, Birla Institute of Scientific Research).

## AN OVERVIEW OF THE IMPORTANCE OF BLEACHING AND THE PAPER INDUSTRY:

In the pulp processing industry, there are two methods: mechanical or chemical. Here, the colorless fibres and lignin are depolymerized are dissolved. Two processes: the sulphite process and the kraft (sulphate) process. The latter process yields a more stronger pulp than other processes with efficient recovery of feedstock chemicals.

Following this step, a bleaching step is used for the manufacture of bright papers the residual fibers following the kraft process. Several chemicals are used in this process of which chlorine is most ubiquitous. The chemical is a powerful electrophile and oxidizing agent to bleach efficiently however, it is associated with certain adverse effects on the environment.

Industries employed around 10% of chlorine in the form of chlorine dioxide were used that is slowly changed to the use of chlorine dioxide more predominantly instead of molecular chlorine chlorine dioxide

The production of pulp involves the separation of cellulose from lignin with great care to minimize the damage to the fibers followed by the separation of lignin from the fibres: a process called delignification.

The process of removal of lignin results in papers with increased brightness however; huge quantities of materials are generated in this process. A 1985 study evaluated the amounts of materials that are generated in the process that include:

100 000 tonnes of particulates

80 000 tonnes of sulphur dioxide

60 000 tonnes of carbon monoxide

15 000 tonnes of nitrogen oxide

12 000 tonnes of hydrocarbons

It was also reported that around 80% of paper mills in Canada release their effluents into water bodies. The amount of total suspended solids (TSS), biological oxygen demand (BOD) and toxic discharge are evaluated of which the latter accounts for failed lethality tests.

The effects of bleaching agents used in the industry, especially chlorine based compounds such as organochlorines and dioxins have been reported (Solomon, 1996).

## AN OVERVIEW OF CHEMICAL BLEACHING AGENTS

Bleach is a chemical that whitens a material. They find application in several aspects ranging from the house to whiten clothes, disinfect floors and bathrooms. They also find use in swimming pools to disinfect water and also in several industries to lighten the color of the products or raw materials used.

A common chemical base for bleach is chlorine used in the form of sodium hypochlorite or calcium hypochlorite. Other common chemicals include sodium percarbonate, hydrogen peroxide that use oxidation against sodium borohydride or sodium dithionite that use the principle of reduction.

History records the use of bleach fields that were used to whiten clothes in the sun and water (Aspin, 1981). Following the discovery of chlorine, Claude Berthollet synthesized Eau de Javel, or Javel water: sodium hypochlorite. Another scientist, Charles Tennant synthesized calcium hypochlorite (Chisholm, 1911).

There are several bleaches that find use in the world:

Chlorine based bleaches use the principle of the amount of active chlorine present in the chemical. The electrolysis of sodium chloride yields chlorine gas as per the following equation:

$2\ NaCl + 2\ H_2O \rightarrow Cl_2 + H_2 + 2\ NaOH$

Chlorine is used to prepare hypochlorite either sodium or calcium. It found extensive use in the bleaching of pulp in the paper industry. However, its use has been replaced by others due to its effect on the environment.

Chlorine dioxide finds use in bleaching of a host of products such as pulp, flour, textiles, etc. It is prepared from chlorine and sodium chlorite.

$2\ NaClO_2 + Cl_2 \rightarrow 2\ ClO_2 + 2\ NaCl$

Research indicates its preparation from sodium chlorate that on reduction with sulfur dioxide or methanol or hydrochloric acid or hydrogen peroxide yields Chlorine dioxide (Vogt, 2010 and Ni et al, 1996)

$2\ NaClO_3 + 2\ HX + \text{“R”} \rightarrow 2\ NaX + 2\ ClO_2 + \text{“RO”} + H_2O$

R is the reducing agent and RO is oxidized form.

Hypochlorite based bleaches:

The hypochlorite along with chlorine in water in equilibrium with each other whose concentrations are affected by pH as shown by the following equation:

$Cl_2 + H_2O \qquad H^+ + Cl^- + HClO$ (Hypochlorite)

According to the pH, there are shifts in the balance between chlorine and hypochlorite.

Mixing hypochlorite bleach with an acid can liberate chlorine gas. Hypochlorite and chlorine are in equilibrium in water; the position of the equilibrium is pH dependent and low pH (acidic) favors chlorine (Cotton and Wilkinson, 1972).

However, chlorine gas has been classified as an irritant of the mucous membrane and respiratory tracts. Occupational Safety & Health Administration (OSHA), US has placed a cap of 0.5 ppm, an average weighed out over a period of 8 hours in a 38-hour week (Occupational Safety & Health Administration, 2007).

The reaction of ammonia with sodium hypochlorite yields several products such as (Rizk-Ouaini et al, 1986) chloramines, dichloramine and nitrogen trichloride. The former two are irritants while the latter is explosive.

The equations are summarized as below:

$NH_3 + NaOCl \rightarrow NaOH + NH_2Cl$

$NH_2Cl + NaOCl \rightarrow NaOH + NHCl_2$

$NHCl_2 + NaOCl \rightarrow NaOH + NCl_3$

**Sodium hypochlorite** is ubiquitously used especially as bleach that is a dilute solution of the chemical in aqueous solution.

Its preparation involves electrolysis of NaCl solution.

$2\ Cl^- \rightarrow Cl_2 + 2\ e^-$

$Cl_2 + H_2O \leftrightarrow HClO + Cl^- + H^+$

Alternatively, its preparation involves the reaction of caustic soda and chlorine gas.

$Cl_2\ (g) + 2\ NaOH\ (aq) \rightarrow NaCl\ (aq) + NaClO\ (aq) + H_2O\ (l)$

(Lenntech.com)

This sodium hypochlorite finds use as a disinfectant in several places and as chlorine gas has certain challenges especially in transportation, the use of sodium hypochlorite is preferred (Intratec).

Bleaching powder or Calcium hypochlorite is referred to as chlorinated lime that has greater stability than sodium hypochlorite.

Apart from chlorine based bleaches, there are peroxide based bleaches that include Hydrogen peroxide which is a powerful oxidizing agent that is used in the bleaching of wood in the paper industry apart from other materials.

A more eco-friendly chemical is Sodium percarbonate produced by reaction between sodium carbonate and hydrogen peroxide as shown in the following equation:

$2Na_2CO_3 + 3H_2O_2 \rightarrow 2Na_2CO_3.3H_2O_2$

Another chemical called Sodium perborate finds its origin in borax ($Na_2B_4O_7$) that is subjected to treatment with caustic soda to yield sodium metaborate ($NaBO_2$) that on treatment with hydrogen peroxide yields the perborate:

$Na_2B_4O_7 + 2\ NaOH \rightarrow 4\ NaBO_2 + H_2O$

$2\ NaBO_2 + 2\ H_2O_2 + 6\ H_2O \rightarrow [NaBO_2(OH)_2\ x\ 3\ H_2O]_2$ (Sodium Perborate REACH Consortium)

Other bleaches employed include ozone that finds huge application in the paper and pulp industry however the use of ozonizers has been attributed to increasing the cost of production (Mojsov et al, 2016).

## EFFECT OF THE USAGE OF CHEMICAL BLEACHING AGENTS ON THE ECOLOGY:

The effluents of pulp mills that involve bleaching stages have been associated with changes in the reproductive and endocrine systems, especially with the amounts of steroids in plasma of fish. It was pointed out that this change could be attributed to a molecule that was formed before the bleaching stage in the processing industry.

Apart from changes in steroid levels of fish, the bleaching agents have been associated with a delay in maturation along with a decrease in GSI: gonadal: somatic indices in studies carried out in fathead minnows.

In studies involving goldfish, the organism displayed a reduction in the levels of steroids in its system when exposed to such effluents. There was reduction in the levels of steroids in fish that were present in the Jackfish Bay, Kapuskasing and Smooth Rock Falls (Canada) (Solomon, 1996).

### Case study: Paper Pulp Effluent and Effect on Ecology

Studies involving the effects of primary treated bleached kraft mill effluent (BKME) were performed in a project called 'EnvironmentKellulose'. The effects of the effluent from Norrsundet mill was studied on fish inhabiting Gulf of Bothnia. It was observed that plants and animals were affected in proportion to the distance that they occupied from the effluent. For instance, there were 8 plant species in the radius of 2km of the effluent while the numbers shot up to 20 in a radius, 4.5 km away from the effluent discharge point.

Fibers from the discharge site were observed that occupied spaces in the bottom of the waters causing a dip in the levels of oxygen. The changes in the levels of such fibres and light along with the presence of chlorine adversely affected *Fucus vesiculosus*, a brown alga.

The adverse effects on fish population were seen in the form of a dip in the diversity as well as numbers of fish less than 2 km from the discharge. As the distance increased in the span of 2 km to 4.5 km from the effluent site, the numbers of adults and fry of *Rutilus rutilus* along with *Gymnocephalus cernua* was observed (Solomon, 1996).

## BIOBLEACHING OF OILS

Apart from the paper industry, biobleaching has been reported in the food industry. The bleaching of vegetable oils removes the color and other components such as peroxides and soaps. The use of charcoal and acid activated clay (AAC or Tonsil Earth) has a drawback of cost. A research team looked at the potential of material of biological origin: rice husk ash and rice husk silica. It was reported that rice husk ash along with carbon and acid treated rice silica were effective in reducing the dependence of costly AAC (De et al, 2009).

## MICROBES AND BIOBLEACHING

Research indicates that the delignification employing fungi can replace 72% of the chemicals needed to bleach a kraft pulp. The use of biological removal of lignin requires milder conditions as compared to that of chemical conditions. Several white rot fungi such as *Phanerochaete chrysosporittm* better than *Trametes* (*Coriolus*) *versicolor, Gloeoporus dichrous*, and *Lentinus edodes* at delignifying softwood kraft pulp. *T. versicolor* was found to give good bleaching effects when compared to be other fungi.

It is interesting to note that several species show variation in their ability to bleach wood. Of several monokaryotic and dikaryotic strains of *T. versicolor* tested, it was found that monokaryons showed better ability to bleach as compared to dikaryons.

A new isolate called IZU-154 was found to bleach hardwood at a better rate as compared to *T. hirsuta*, *T. versicolor*, and *P. chrysosporium*.

Research also indicates that the use of biological agents to bleach wood showed good results and in combination with peroxide a good effect was seen in order to replace the use of chlorine based bleaching agents.

Apart from microbes several enzymes have been studied for the purpose of bleaching that include:

Laccases of several isoforms of fungi such as *T. versicolor*, could bleach pulp in combination with certain compounds such as 2,2'-azinobis(3-ethylbenzthiazoline-6-sulfonate) (ABTS) to bleach wood.

The other two main enzymes studied include Manganese and lignin peroxidases that were detected from several white rot fungal species.

Another enzyme, Cellobiose quinone oxidoreductase (CBQase) was isolated from *T. versicolor* that was attributed with bleaching activities. The enzyme targets several products of the outcome of the action of laccases or peroxidases (Reid and Paice, 1994).

## A GLIMPSE OF MICROBES USED FOR BIOBLEACHING

Several microbes have found use in the bleaching industry that are summarized in the following table: (Archna et al, 2012)

| Sr. No. | Dye targeted | Microbe | Reference |
|---|---|---|---|
| 1. | Acidblue113 | *Bacillus subtilis* | Gurulakshmi et al, 2008 |
| | Reactive red 180 | *Citrobacter sp. Ck3* | Hui wang et al, 2009 |
| | Several industrial dyes such as Tropaeolin O, Reactive Blue15, RemazolOrange | *White rot fungi such as Bjerkandera sp, Phanerochaete chrysosporium, Trametes hirsute, Trametes versicolor, Pleurotus ostreatus* | Swamy and Ramsay, 1999 |
| | Azo dye- Red 3BN | *Penicillium chrysogenum* | Kumar Praveen, 2012 |
| | Reactive yellow 145 | *Enterococcus faecalis strain YZ66* | Sahasrabudhe, 2011 |
| | azo dye | *Listeria sp* | Kuberan, 2011 |
| | Textile dye: orange 3R | *Klebsiella sp. Salmonella sp. Pseudomonas sp.* | Ponraj, 2011 |

## SUMMARY

The process of bleaching finds application in a wide range of industries. There are several methods of removing color or dyes that involve physical, chemical or biological methods. The use of biological methods shows several advantages such as economic cost and the biodegradable nature of the compounds that are used.

Several microbes, especially white rot fungi have found a lot of application in the process of biobleaching. The enzymes of the microbes that have been studied include laccases, xylanases, manganese peroxidases and lignin peroxidases. The use of the enzymes and microbes has shown good results in the bleaching of several recalcitrant dyes. Enzymes and microbes have been isolated from soil, wood and even vegetable sources such as garlic. (Discussed in Chapter 2 and Chapter 4).

Further research and studies are warranted in order to enhance the field of biobleaching as it offers a safer and more viable alternative and minimizing the damage to the already fragile environment.

## REFERENCES

1. Archna, Lokesh KN, Kiran SRR (2012) Biological methods of dye removal from textile effluents - A review. J Biochem Tech 3(5): S177-S180.
2. Aspin C (1981) The Cotton Industry. Shire Publications Ltd. p. 24. ISBN 0-85263-545-1
3. Bijay Krishna BD, Patel JD, Patel JB, et al (2009) Bleaching of mustard oil with alternative bleaching agents and acid activated clay. J Oleo Sci. 58 (2): 57-63.
4. Chisholm H, ed. (1911). Bleaching. Encyclopædia Britannica (11th ed). Cambridge University Press.
5. Cotton FA, Wilkinson G (1972) Advanced Inorganic Chemistry. John Wiley and Sons Inc. ISBN 0-471-17560-9.
6. Gurulakshmi M, Sudarmani DNP, Venba R, et al (2008) Biodegradation of leather acid dye by Bacillus subtilis. Adv Biotech 12:12-20.
7. Wang H, Su QJ, Zheng XW, et al (2009) Bacterial decolorization and degradation of the reactive dye reactive red 180 by Citrobacter

sp. Ck3. Int Biodeterioration Biodegradation 63:395–399.

8. Reid ID, Paice MG (1994) Biological bleaching of kraft pulp by white rot fungi and their enzymes. FEMS Microbiol Rev 13: 369-376.
9. Solomon KR (1996) Chlorine in the bleaching of pulp and paper. Pure Appl Chem 68: 1721-1730.
10. Mojsov KD, Andronikov D, Aco Janevski A, Kuzelov A, Gaber S (2016) The application of enzymes for removal of dyes from textile effluents. Advanced Technologies. 5(1): 36-41
11. Koponen R (1991) Enzyme systems prove their potential. Pulp pap Int 33:81–83.
12. Kuberan T, Anburaj JJ, Sundaravadivelan C, et al (2011) Biodegradation of azo dye by Listeria sp. Int J Env Sci 1:1760-1770.
13. Kumar PGN, Sumangala K Bhat SK (2012) Fungal Degradation of Azo dye- Red 3BN and Optimization of Physico-Chemical Parameters ISCA J Biol Sci 1:17-24.
14. Occupational Safety & Health Administration (2007) OSHA. Retrieved 2007-08-26.
15. Ghosh P, Choudhary B, Mantri R, et al (2002) Biobleaching: Enzymatic colour removal from pulp. Final report submitted to Indian Paper Manufactures Association, New Delhi. Birla Institute of Scientific Research Statute Circle, Jaipur.
16. Ponraj MK, Gokilal, Zambare V, et al (2011) Bacterial decolorization of textile dye- orange 3R. Int J Adv Biotech Res 2:168-177.
17. -Rosette R-Q, Michel F, Josette G, Adad S-C, Therese M (1986) Oxidation reaction of ammonia with sodium hypochlorite. Production and degradation reactions of chloramines. Bull Soc Chim Fr 4: 512– 521.
18. Sahasrabudhe MM, Pathade GR (2011) Decolourization of C.I. reactive yellow 145 by Enterococcus faecalis strain YZ66. Arc Appl Sci Res 3:403-414.
19. Archana A, Sharma A, Satyanarayana T (1997) Xylanolytic enzymes. The Botanica 47:163-167.
20. Sodium hypochlorite as a disinfectant. Lenntech.com. Retrieved 2011-08-07.

21. Sodium Hypochlorite Chemical Production. Intratec, ISBN 978-0615702179.
22. Sodium Perborate REACH Consortium. Retrieved 2012- 06-07.
23. Swamy J, Ramsay JA (1999) The evaluation of white rot fungi in the decoloration of textile dyes. Enz Micro Tech 24:130–137.
24. Viikari L, Kantelinen A, Sundquist J, Linko M (1994) Xylanase in bleaching: from an idea to the industry. FEMS Microbiol Rev.13:335–350.
25. Viikari L, Ranuae M, Kantelinen A, Linko M, Sundquist J (1986) Application of enzymes in bleaching. Proceedings of the 3rd international conference on biotechnology in the pulp and paper industry, Stockholim, pp 67–69.
26. Vogt H, Balej J, Bennett J E, Wintzer P, Sheikh S A, Gallone P, Vasudevan S, Pelin K (2010) Chlorine Oxides and Chlorine Oxygen Acids. Ullmann's Encyclopedia of Industrial Chemistry. Wiley-VCH. doi:10.1002/14356007.a06_483.pub2.
27. Ni Y, Wang X (1996) Mechanism of the Methanol Based ClO2 Generation Process. International Pulp Bleaching Conference. TAPPI. pp 454–462.

# CHAPTER 2

# ENZYMES / MICROBES USED FOR BIOBLEACHING

**Shiv Sanjeevi Sripathi[1] and Prerna Pandey[2]**

[1]Shiv Sanjeevi Sripathi completed his BSc in Biotechnology from Mumbai University and secured second rank in Mumbai University and first rank in his college: Kishinchand chellaram College. Shiv secured first rank in his college KET's V.G.Vaze College and has worked on a stem cell project at the Specialized Centre for Cell Based Therapy (SCCT), KEM Hospital at Mumbai, following which he was a Junior Research Fellow at Microbiology & Cell Biology Department, Indian Institute of Science, Bangalore from August 2009 to November 2011. The field of work involved cloning of cell wall genes and transcription factors in E. coli & M. smegmatis. This was followed by a stint as a Research and Development Executive at Transasia Biomedicals, Mumbai where he worked on patent writing and developing immunoassays from 2011 till 2013. In 2013, he entered the teaching field where he has been tutoring biology and chemistry to the students of 11th and 12th grades along with English grammar and communication skills to people of all age groups.

[2]Dr Prerna Pandey, a biotechnologist with several years of wet lab research experience, worked at International Center for Genetic Engineering and Biotechnology, New Delhi. Her field of research in PhD included isolation and molecular characterization of geminiviruses, genome sequencing, gene annotation, and gene silencing using the RNA interference technology. She has also worked at Transasia Biomedicals and Advance Enzyme Technologies as a scientist. Prerna has published papers in peer-reviewed journals, and has submitted a number of annotated Geminiviral genome sequences in the GenBank, including two novel ones. She has also completed her editing and proof-reading courses from Society for the promotion of Editors and Proof-readers, UK and now works as a freelance scientific editor and writer. When Prerna is not busy with her assignments, she loves to explore historical places.

Several chemicals used in conventional bleaching include peroxidises, sodium hypochlorite, urea etc. that are associated with damage to the containers and also the evolution of pollution. The bleaching agent depends on the textile, its handling and whether it requires a bleaching agent (Ramadan, 2008).

## DYES

Dyes are defined as recalcitrant organic molecules that resist to aerobic digestion and are stable to light (Archna et al, 2012). It yields color to several products like foodstuff, paper and textiles. Synthetic dyes are classified into several groups such as nitro, azo, anthroquinones etc. These synthetic dyes are non-biodegradable and hence exert toxic effects on the environment (Murugesan and Kalaichelvan, 2003).

The azo group of dyes have a functional group R-N=N-R' (R and R' are aryl or alkyl) and are reported to be resistant to oxidative degradation processes (Archna et al, 2012). These dyes represent one of the largest category used in the printing and textile processing industries. These dyes and their aromatic amine products produced on degradation are toxic (Brown and Hamburger, 1987).

## METHODS OF BLEACHING:

The methods of bleaching include physical, chemical and biological methods:

### Physical Methods

These include methods such as filtration using membranes and adsorption. The use of membranes has can be integrated in water treatment plants when compared to other systems where a separate treatment step is required.

Ultrafiltration has been used in bioreactors or prior to reverse osmosis treatments. It has also found use in the removal of hydrophobic spin finish compounds from textile wastes.

Microfiltration has found use in the treatment of dye baths that have the dye and rinsing washes. It has also been used prior to other filtration or reverse osmosis treatments.

Reverse osmosis aids in the separation of salts, dyes that have been subjected to hydrolysis and boast of a high retention value of the membranes used although, the techniques are associated with higher power usage.

Adsorption techniques find use in the removal of dyes from solution and necessitate the low cost of the absorbent and the adsorbent not dependent on pre-treatment steps. Substances such as carbon need a longer time duration for the removal of such dyes and incur higher expenditures.

## Chemical Methods

There are several chemical techniques used for bleaching that are coagulation or flocculation along with flotation and filtration, flotation using electric means, oxidation using oxidizing agents, radiation, $Fe^{2+}$/ $Ca(OH)_2$ that cause precipitation. However, the chemicals used are found to add to the cost of the process and also form a slurry that requires additional removal.

The use of ozone as an oxidizing agent has reported high levels of bleaching however, the technique entails the use of costly and separate system.

Fenton's Oxidation and electro chemical methods of oxidation have also been reported as good bleaching techniques but are limited by the formation of sludges or slurries that require additional removal.

The technique of coagulation finds application in the removal of soluble dyes. The use of gums from certain plants have reported better bleaching action than chemical coagulating agents. Such gums include guar gum or extracts from *Ipomoea dasysperma* have yielded good bleaching effects (Archna et al, 2012).

These conventional methods discussed above generate a large quantity of sludge, they also need the involvement of an external agent, increase the associated costs and also are long drawn protocols that are associated with operational tediousness (Rodr´ıguez et al, 2010).

## Biological Methods of Bleaching

The use of biological techniques includes the use of bacteria, fungi and algae for the degradation of dyes.

The use of microbial cells or biomass: viable or dead have been used for various effluents that include aerobic and anaerobic systems. Aerobic systems entail the use of several aerobic microbes like bacteria to degrade several dyes. However, the azo dyes have shown resistance to the aerobic treatment systems. The anaerobic systems use bacteria of several; species such as Eubacterium, Clostridium and Bacteroides. These organisms generally lyse the azo bond to yield several amine products. Additionally, fungi can be used in several systems such as bioreactors for the bleaching of dyes (Archna et al, 2012; Kiro et al, 2016).

Certain studies have shown that the use of bacteria in bleaching of dyes especially few azo dyes were not up to the mark. When the dyes were subjected to anaerobic degradation, certain bleached aromatic amines were formed that were found to exert toxic effects like mutagens or carcinogens (Isik and Sponza, 2007).

Biological treatments are pitched as economic as compared to physical or chemical methods. Microbes also can perform the concentration and degradation of several complex molecules.

## USE OF ENZYMES IN BIOBLEACHING

The use of enzymes in the process of biobleaching finds good potential in the removal of commercial dyes that include azo, anthraquinone, phthalocyanine, triphenylmethane and heterocyclic compounds. Such groups were found to be effectively degraded by several enzymes produced predominantly by white rot fungi.

Advantages of the usage of enzymes in the field of biobleaching (Kiro et al, 2016) include:

- The catalysts are biodegradable proteins.
- Enzymes are highly specific.
- They can be operated over a broad range of conditions such as temperature, pH and salt.
- They can function across a wide range of concentrations of substrates.
- The formation of sludge is lesser when compared to other catalysts or microbes.

- The developments in technology and bioreactors offer ease of manipulation and control.
- With the advent of enzyme engineering techniques, enzymes can be designed through the application of bioinformatics to suit a particular purpose.

# ENZYMES USED

Three types of enzymes have been described produced by white rot fungi that include benzene diol oxidoreductases or laccases (1.10.3.2), manganese dependent peroxidases (1.11.1.13) and lignin peroxidases (LiPs, EC 1.11.1.14) (Duenas and Martinez, 2009).

These enzymes have been reported to be produced by the fungi under conditions of nutrient limitation such as nitrogen, carbon or sulfur (Barr and Aust,1994).

The modification of lignin can be achieved by the two peroxidases called lignin peroxidase and manganese peroxidase along with the enzyme laccase, an oxidase.

## Lignin Peroxidase

The enzyme catalyzes the oxidation of a phenol group to produce a radical at the carbon attached to the azo linkage. This phenolic carbon reacts with water to produce a phenyldiazene group that is oxidized using an electron to yield Nitrogen.

## Manganese Peroxidase

The enzyme has been attributed with high potential for degradation. It catalyzes the oxidation of Mn2+ to Mn3+ ions causing a spurt in its reactivity. The azo groups are broken down into quinines, organic acids and carbon dioxide that can be acted upon by other systems.

## Laccase

This enzyme removes an H+ atom from hydroxyl and amino groups of the ortho- and para- substituted mono or polyphenolic substrates apart from aromatic amines. Apart from bleaching the enzyme finds use in the removal of chloro-lignins or phenolic groups such as textile,paper and pulp, tannery and effluents (Reviewed by Kiro et al, 2016).

The enzyme laccase is described as a copper containing protein that has been reported as effective in the degradation of dyes especially azo dyes. The enzyme is produced in an extracellular fashion and is classified as a blue oxidase. The enzyme acts using the mechanism of oxidation rather than other systems such as targeting certain groups.

The advantages associated with laccase include:

- It is non- specific and hence can be used for the bleaching of several dyes.
- It does not cause any effect on the environment.
- It does not depend on a cofactor for its functioning simplifying the process.
- It is produced in an extracellular fashion that adds to the simplicity of purification.
- (Archna et al, 2012).
- The specificity of the enzyme was shown to be further boosted by the usage of certain redox compounds produced by lignin degrading fungi or chemicals such as ABTS (2,2'-azino-bis(3-ethylbenzthiazoline) sulphonic acid), 1-hydroxybenzotriazole (Ledakowicz et al. 2001).

Laccases have been shown to be present in several organisms such as bacteria, white rot fungi and insects. Literature has explored the use of laccases from several sources especially fungi such as *Phanerochaere chrysosporium ,Pleurotus ostreatus, Trametes versicolor, Phlebia tremell , Pyricularia oryzae, Pycnoporus cinnabarinus* .

## USE OF ENZYME LACCASE IN BIOBLEACHING

The use of the enzyme in direct form rather than the usage of entire microbial cells have shown higher efficiency. A 2004 study showed that laccases purified from the organisms *Trametes hirsute* and *Sclerotium rolfsii* showed degradation of azo dyes. The enzymes showed enhanced activity in the presence of a redox compound acetosyringone (Kandelbauer et al, 2004).

In another approach, a research team in 2008 worked on the development of a laccase based system along with mediators to bleach dyes in effluent waters. The authors tested several mediators such as (2,2-azinobis(3-ethylbenzothiazoline-6-sulfonate) (ABTS), 1-hydroxybenzotriazole (HBT), N-hydroxyacetanilide (NHA), polioxometalates, violuric acid (VA) and (2,2,6,6-tetramethylpiperidin-1-yloxy) (TEMPO) on several dyes used in textiles.

They reported a bleaching of 70% when ABTS was used with laccase at a prescribed temperature and pH on the dyes reactive black 5, reactive blue 114 and reactive yellow 15. A 93% decolorization was achieved with ABTS on reactive blue 114 (Tavares et al, 2008).

## MECHANISMS OF THE ENZYMES USED

The mechanisms of biobleaching by lignin peroxidases and manganese peroxidases have been attributed to oxidation. The phenol groups are oxidized to yield a free radical at the carbon attached to the azo moiety that reacts with water to yield a phenyl diazine that yields Nitrogen.

The enzyme laccase has gained much importance in the field of bleaching. The mechanism used by the enzyme includes oxidation to first yield a phenoxy ion followed by a carbonium ion. This reacts with water to form quinone and a diazene derivative (Murugesan and Kalaichelvan, 2003).

## AZO-REDUCTASE

The extra cellular enzyme converts azo compounds to aromatic amine derivatives utilizing NADPH. The enzyme has been reported as non-specific in the degradation of azo compounds of sewage sludge. The enzyme also has reported high success values with the majority of reports in literature showing the reduction of azo compounds to amine derivatives.

The use of azo reductases from aerobic systems such as Pseudomonas strains K22 and KF46 have been reported as effective in biobleaching. These enzymes work as monomers and utilize NAD/NADH and use the mechanism of reduction to bleach sulfonated azo drugs.

The bacterial strains that have been reported as effective include *Bacillus subtilis, Acinetobacter sp, Aeromonas hydrophilam, Pseudomonas luteola, Acetobacter liquefaciens S-1 and B. cereus*. These organisms have received special mention as these organisms can carry out bleaching in the conditions of the wastewater especially in terms of temperature and pH (Archna et al, 2012).

## LINEN AND GLUCOSE OXIDASE

In an attempt to examine the use of enzymes in the bleaching of textiles, an article was published that dealt with the use of an enzyme, glucose oxidase that oxidizes glucose as a substrate to release hydrogen peroxide. With growing concerns about wastewater management and the side effects of conventional agents used as bleach, the author(s) reported the use of a biological approach.

It was reported that linen and its blends needed the use of special treatments as the levels of impurities are high. Three different types of textiles were used that included cotton/linen in several ratios and also pure cotton. They were subjected to conventional chemical treatment that involved the use of a mixture of sodium hydroxide, sodium silicate, sodium carbonate, magnesium sulfate and hydrogen peroxide. The process of biobleaching involved the use of glucose oxidase and its substrate glucose.

Following treatment, it was observed that the tensile strength of the fabric was higher in the case of biobleaching than conventional chemical bleaching. The reduction in the tensile of the fabric was reported at 11-14% for chemical bleach while a value of 4-7% for that of biobleaching. The author(s) observed that the level of whiteness in the fabric increased up to a level of 25U following which there was a decline. The level of whiteness was reported as marginally more in the chemical treatment.

The presence of glucose in the treatment solution was found to enhance the whiteness of the fabric. When the aspect of time was considered, samples showed a good degree of whiteness to chemical bleaching below 90 minutes while biobleaching showed comparable results at 90 minutes. The effect of biobleaching on the structure of cotton fibers revealed that while the untreated fibers had parallel ridges, the biobleached samples showed the presence of more smoothened and polished surfaces.

It was overall concluded that the effect of biobleaching was comparable to that of conventional chemical bleaching albeit at a higher cost involved. However, the author(s) highlighted that the use of enzymes is more preferable as it minimizes damage to the environment and there are additional costs with chemical bleaching that entails the use of pollutant neutralizers and associated health effects (Ramadan, 2008).

Another enzyme that has been reported is **xylanase** that is discussed in chapter 4,"Research conducted and case studies: Enzymes from sewage to the kitchen".

# USE OF MICROBES FOR BIOBLEACHING

## Algae and Biobleaching

A 2008 study evaluated the use of algae in the biobleaching of azo dyes such as Tartrazine: monoazo dye and Ponceau : diazo dye. Fast removal of color was achieved in the initial three days of culture that slowly decelerated with time. The algae *S. bijugatus* caused a reduction of upto 68% of the dyes while another alga *N. muscourm* showed decolorization of the dyes. The mechanism of biobleaching by algae is catalyzed by the enzyme Azo reductase that lyses the azo linkage of the dyes (Omar, 2008).

## White Rot Fungi and their Enzymes

The white rot fungi are a heterogeneous group of organisms that degrade fungi and other lignin containing substances. They are named so as the color of delignified wood is whitish: these organisms are the ones capable of degrading the huge and hard to digest lignin. Research has indicated that the ratios of carbon and nitrogen especially nitrogen as a limiting factor plays a major role in the digestion of lignin by virtue of secondary metabolism.

In majority of white-rot fungi synthesize at least two of the three like **lignin peroxidase**, **manganese peroxidise** and **laccases** to yield free radicals during the reaction. There are various factors that influence the bleaching process by the microbes that include: the dye structure, medium used, pH, the conditions of aeration and mixing (Senthil et al, 2013).

The microbes are able to oxidize a vast spectrum of different compounds, and therefore they have been intensively studied in the treatment of wastewater from textiles (Kiro et al, 2016).

These white rot fungi can show a preference for hardwood tress and are quite common in nature. The non-selective white rot fungi can degrade cellulose, hemicelluloses and lignin portions of wood, while the selective white rot fungi can mainly digest lignin.

Advantages of white rot fungi in biobleaching:

- The enzymes produced by the microbes are non-specific
- As these enzymes are produced outside of the body, they can be used for the targeting of a wide variety of chemicals such as dyes.

Pulping entails the removal of lignin from cellulose for paper production. When chemicals are used, for removing lignin, a colored precipitate is produced that requires the use of bleaching agents. The term biopulping is used for the administration of such microbial agents on wood pieces instead of traditional methods such as chlorine.

The use of white rot fungi dates back to 1957 by Lawson and Still at the West Virginia Pulp and Paper Company Research Laboratory now Westvaco Corp (Akhtar et al, 1998).

The use of *Ceriporiopsis subvermispora* has been found to be very useful with processes optimized for the same.

- A few of the merits of the usage of white rot fungi include:
- The ability to withstand elevated levels of pollutants or even toxic xenobiotics as they possess an extra cellular system of enzymes.
- Apart from not requiring a pre-treatment with conditioning system for a pollutant, they can be grown on a host of materials (Reviewed by Rodríguez-Couto S, 2017).
- Certain white rot fungi were also shown to produce additional chemicals that can serve as added byproduct. For instance, *Ceriporiopsis subvermispora* was found to produce a xylanase enzyme that finds use in biobleaching and degradation of wood when the organism was grown on pine (Milagres et al, 2005).

## Azo Dyes and White Rot Fungi

The organism *Phanerochaere chrysosporium* belonging to the group of basidiomycetes was found to degrade several dyes especially polymers (Glen and Gold, 1983). Following this, azo, anthroquinone and other polymers were found to be degraded by the organism.

Research work in 1990 showed that this organism could also degrade several dyes such as orange II, congo red and tropeolin O. Cultures could degrade up to 90% of the dyes in a week of culture Cripps et al, 1990.

Another research study has shown that *Trametes versicolor* was found to bleach and degrade dyes such as flourene, dichloro aniline and benzene pyrene (Murugesan and Kalaichelvan, 2003).

*Trametes versicolor* in the form of pellets was used for bleaching of the textile dye, Grey Lanaset G. High amount of laccase enzymes was detected and the bleaching of the dye was seen to a level of 90% (Blanquez et al, 2006).

A 1997 study showed that apart from *Phanerochaere chrysosporium*, two other species: *Trametes versicolor* and *Bjerkandera adusta* could degrade several commercially used dyes of several fungal species tested. The dyes included reactive orange, reactive black, reactive blues 15 & 38 and reactive violet. 60% of the dye HRB8 was degraded by *Trametes versicolor* and *Bjerkandera adusta* and upto more than 90% of the dye was bleached in 4 days (Heinfling et al 1997).

*P. chrysosporium* has emerged as a good system in bioremediation of textile and pulp and paper mill effluent. It has the capacity to degrade materials of the effluents such as starch, cellulose, pectin, lignin, lignocelluloses. *P. chrysosporium* can also be used to bleach the direct wastewater containing dyes with complex structures. 3 white rot fungi: *Pleurotus ostreatus* (*P. ostreatus*), *Coriolus versicolor* (*C. versicolor*) and *Funalia trogii* (*F. trogii*) were tested in 2007 for their ability to bleach Remazol Brillant Blue Royal (RBBR) and Drimaren Blue CL-BR (DB). It was reported that laccase enzyme played a major role in the bleaching of the dyes especially by *F. trogii* (Emrah et al, 2007).

A more recent study in 2014 looked at the use of the basidiomycete in the degradation of Amido black 10B. They found that up to 98% of the dye is bleached by the organism *P. chrysosporium* by day 3 of culture and action. The ratios of the various nutrients such as ions, glucose and also certain inducers such as starch have been reported in enhancing the process of biobleaching. It was also reported that such a setup could be applied for a bioreactor system (Senthilkumar, et al, 2014).

These fungi use mechanisms of oxidation for the degradation of the dyes in comparison to the bacterial systems that rely on reduction. By virtue of oxidation, *P. chrysosporium* also aids in the removal of toxicity apart from its bleaching action.

Submerged liquid cultures of laccase producing organisms have caused efficient bleaching of triphenyl methane dyes such as malachite green and bromophenol blue. It has been reported that biobleaching can use laccase as the enzyme does not require stringent conditions as with the lignin peroxidases and manganese peroxidases. Another advantage is the copious amount of enzyme that is produced (Murugesan and Kalaichelvan, 2003)

A 2006 study examined the use of a marine fungal isolate of the white rot family. The isolate called, NIOCC # 2a produced mainly laccase followed by manganese peroxidase that bleached several synthetic dyes found in textile, paper and pulp mill and distillery waste. Apart from laccase, there were several other Extracellular polymeric substances (EPS) that aided bleaching (D'Souza et al, 2006).

## BIOREACTOR SYSTEMS FOR BIOBLEACHING

Several bioreactor systems have been tested in the field of biobleaching including bed reactors, rotating systems and film reactors. A 1996 study showed that a dye called Red 553 was bleached using by ligninase in a fixed film bioreactor. The rate of bleaching was above 80% over a period of upto 20 days using this continuous fermentation system (Young and Yu, 1997).

## IMMOBILIZED CULTURES FOR BIOBLEACHING

The technique of immobilization of the cells can have an advantage over the use of free cells. One advantage is the easy scaling up of the system and another advantage is the ease of operations under aseptic conditions. Various materials have been used in the immobilization of organisms such as nylon webs, sintered glasses, polyurethane foams, silicon tubings and porous polystyrene of which foam based matrices are economical. Other materials also include corn cobs, wheat bran pellets, nylon, pine and lignocelluloses (Murugesan and Kalaichelvan, 2003).

The use of immobilized cultures has yielded cultures that last longer and also bleached dyes. A 2003 study involved the immobilization of the fungus *Irpex lacteus* in a pine and plyurethane caused more effective bleaching as compared to free cultures (Kashinath et al, 2003).

A 2013 study compared the utility of immobilized cells against free ones. Apart from free cells that were used for their ability to bleach Amido black B and Procion blue-2G, cells were also immobilized. The cells were encapsulated in polyvinyl alcohol (PVA) and sodium alginate in CaCl2 solution. It was observed that efficient bleaching of the dyes was achieved in both conditions, thus serving as a useful method of biobleaching. The authors sttributed the role of the enzyme laccase, lignin peroxidases and manganese peroxidases for the bleaching action of the microbes (Kumar et al, 2013).

## FACTORS ASSOCIATED WITH BIOBLEACHING

A main factor has been reported as agitation. The mixing of the medium offers an advantage of the fungi forming pellets rather than mats to facilitate optimum mixing of air and nutrients. More satisfactory decolorization has been reported from that of agitated cultures as compared to static culture conditions. Another factor attributed is pH as dyes cause a shift in pH. Overall research indicates standardization of culture conditions in order to suit each organism used to achieve maximum bleaching of dyes (Murugesan and Kalaichelvan, 2003)

## COMPLEX PELLETS AND BIOBLEACHING

A 2000 study used the white rot fungus *Trametes versicolor,* in the form of complex pellets along with activated carbon. The pellets were constructed to have a central carbon core surrounded by the fungus and were found to bleach the dye acid violet 7 more effectively than the fungus or carbon alone suggestive of the symbiotic action of both the agents of bleaching (Zhang and Yu, 2000).

## CONCLUSION

The use of biobleaching techniques involves the use of biomass or enzymes of several microbes like bacteria, white rot fungi and algae. Several enzymes such as azo reductases, lignin peroxidases, manganese peroxidases and laccases have been described. Another enzyme xylanase is described in detail in Chapter 4. These enzymes have been used as such or along with their producers for the bleaching of several dyes. Such systems have demonstrated high levels of decolorization of several dyes including azo dyes that are most hard to degrade.

Further research is required for the optimization of the systems that use biobleaching for their efficient usage in the removal of toxic dyes that apart from adding visual pollution exert toxic effects on the ecology.

## REFERENCES

1. Akhtar M, Blanchette RA, Myers GC, Kirk TKC (1998) An overview of biochemical pulping. Environmentally friendly tech. Young R, Akhtar M (eds.), John Wiley, New York, pp309-419.
2. Archna, Lokesh KN, Kiran SRR (2012) Biological methods of dye removal from textile effluents - A review. J Biochem Tech 3(5): 177-180.
3. Barr DP, Aust SD (1994) Mechanisms white rot fungi use to degrade pollutants. Environ Science Technol 28:78-87.
4. Blanquez P, Caminal G, Sarra M et al (2006) Study of the cellular retention time and the partial biomass renovation in a fungal decolourisation continuous process. Water Res 40:1650-1656.
5. Brown D, Hamburger B (1987) The degradation of dyestuffs, Part 111 – investigations of their ultimate degradability. Chemosphere16: 1539-1553.

6. Cripps C, Bumpus J, Aust SD (1990) Biodegradation of azo and heterocyclic dyes by Phanerochaere chrysosporium. Appli Environ Microbiol 56: 1114.
7. D'Souza DT, Tiwari R, Sah AK, et al (2006) Enhanced production of laccase by a marine fungus during treatment of colored effluents and synthetic dyes. Enz Micro Tech 38:504- 511.
8. Duquesne S, Bordes F, Fudalej F, Nicaud JM, Marty A (2012) The yeast Yarrowia lipolytica as a generic tool for molecular evolution of enzymes. Methods Mol Biol 861: 301–312.
9. Emrah AE, Ali Ü, Halil K (2007) Decolorization of synthetic dyes by white rot fungi, involving laccase enzyme in the process. Process Biochem 42(10): 1429-1435.
10. Glen JK, Gold MH (1983) Decolorization of several polymeric dyes by the lignin degrading basidiomycete Phanerochaere chrysosporium. Appli Environ Microbiol 45: 1741.
11. Heinfling A, Bergbauer M, Szewyzk (1997) Biodegradation of azo and pthalcocyanine dyes by Trametes versicolor and Bjerkandera adusta. Appli Environ Microbiol 48: 261.
12. Isik M, Sponza DT (2007) Fate and toxicity of azo dye metabolites under batch long-term anaerobic incubations. Enzyme Microb Technol 40: 934-939.
13. Kandelbauer A, Erlacher A, Cavaco-Paulo A, Guebitz M, et al (2004) Laccase-catalyzed decolorization of the synthetic azo-dye Diamond Black PV 200 and of some structurally related derivatives. Biocat Biotrans 22:331-339.
14. Kashinath A, Novotny CK, Kamalesh KS, Patel C, et al (2003) Decolourization of synthetic dyes by Irpex lacteus in liquid culture and packed bed bioreactor. Enzyme Microb Technol 32: 167.
15. Kiro MD, Andronikov D, Janevski A, Kuzelov A, Gaber S (2016) The applications of enzymes for the removal of dyes from textile effluents. Advanced Technologies. 5(1):36-41.
16. Ledakowicz S, Solecka M, Zylla R, et al (2001) Biodegradation, decolourisation and detoxification of textile wastewater enhanced by advanced oxidation processes. J Biotechnol 89:175-184.

17. Milagres AMF, Magalhaes PO, Ferraz A (2005) Purification and properties of a xylanase from Ceriporiopsis subvermispora cultivated on Pinus taeda. FEMS Microbiol Lett 253: 267-272.

18. Murugesan K, Kalaichelvan PT (2003) Synthetic dye decolourization by white rot fungi. Indian J Exp Biol 41: 1076-1087.

19. Omar HH (2008) Algal decolorization and degradation of monoazo and diazo dyes. Pak J Biol Sci 11(10):1310-6.

20. Ramadan AR (2008) Characterization of biobleaching of cotton/ linen fabrics. Journal of Textile and apparel, Technology and Management 6(1): 1-12.

21. Rodr´ıguez G, Ovejero JL, Sotelo M, Mestanza, J Garc´ıa (2010) Heterogeneous fenton catalyst supports screening for mono azo dye degradation in contaminated wastewaters. Ind Eng Chem Res 49(2): 498–505.

22. Rodríguez-Couto S (2017) Industrial and environmental applications of white-rot fungi. Mycosphere 8(3):456–466

23. Ruiz Dueñas FJ, Martínez AT (2009) Microbial degradation of lignin: how a bulky recalcitrant recalcitrant polymer is efficiently recycled in nature and how we can take advantage of this. Microbial Biotechnol 2: 164 -177.

24. SenthilKumar S, Balasubramanian P, Swaminathan G (2013) Degradation potential of free and immobilized cells of white rot fungus Phanerochaete chrysosporium on synthetic dyes. International Journal of ChemTech Research Coden (USA) 5(2): 565-71 IJCRGG ISSN: 0974-4290.

25. Senthilkumar S, Perumalsamy H. Janardhana Prabhu (2014) Decolourization potential of white-rot fungus Phanerochaete chrysosporium on synthetic dye bath effluent containing Amido black 10B. J Saudi Chem Soc 18(6): 845-853.

26. Tavares A, Cristóvaão R, Lourerio J, et al (2008) Optimization of reaction textile dyes degradation by laccase-mediator system. J Chem Tech Biotech 83:1609-1615.

27. Young L, Yu J (1997) Ligninase catalyzed decolourization of synthetic dyes. Water Res 31: 1187.

28. Zhang F, Yu J (2000) Decolourization of acid violet 7 using complex pellets of white rot fungus and activated carbon. Bioprocess Engg 23:295.

# CHAPTER 3

# TRADITIONAL BLEACHING AND BIOBLEAHING

**Shiv Sanjeevi Sripathi[1] and Prerna Pandey[2]**

[1]Shiv Sanjeevi Sripathi completed his BSc in Biotechnology from Mumbai University and secured second rank in Mumbai University and first rank in his college: Kishinchand chellaram College. Shiv secured first rank in his college KET's V.G.Vaze College and has worked on a stem cell project at the Specialized Centre for Cell Based Therapy (SCCT), KEM Hospital at Mumbai, following which he was a Junior Research Fellow at Microbiology & Cell Biology Department, Indian Institute of Science, Bangalore from August 2009 to November 2011. The field of work involved cloning of cell wall genes and transcription factors in E. coli & M. smegmatis. This was followed by a stint as a Research and Development Executive at Transasia Biomedicals, Mumbai where he worked on patent writing and developing immunoassays from 2011 till 2013. In 2013, he entered the teaching field where he has been tutoring biology and chemistry to the students of 11th and 12th grades along with English grammar and communication skills to people of all age groups.

[2]Dr Prerna Pandey, a biotechnologist with several years of wet lab research experience, worked at International Center for Genetic Engineering and Biotechnology, New Delhi. Her field of research in PhD included isolation and molecular characterization of geminiviruses, genome sequencing, gene annotation, and gene silencing using the RNA interference technology. She has also worked at Transasia Biomedicals and Advance Enzyme Technologies as a scientist. Prerna has published papers in peer-reviewed journals, and has submitted a number of annotated Geminiviral genome sequences in the GenBank, including two novel ones. She has also completed her editing and proof-reading courses from Society for the promotion of Editors and Proof-readers, UK and now works as a freelance scientific editor and writer. When Prerna is not busy with her assignments, she loves to explore historical places.

# INTRODUCTION

For a material to be perfectly colored, all kinds of impurities, natural or acquired, need to be removed for the colorants to either adjust perfectly on the surface or penetrate inside the substrate, for the colorant to be clearly visible without any interference.

Before the textile is dyed, a number of preparatory processes are performed. The traditional processes used to prepare cotton fiber, yarn or fabric have the most negative environmental impacts. For example, the conventional preparation of cotton for dyeing is highly alkaline, which account for nearly 75% of the organic pollutants arising from textile finishing (Hossain and Uddin, 2011). For the conventional preparation, sodium hypochlorite solutions, or concentrated sodium hydroxide solution and hydrogen peroxide are applied on raw cotton for removing the impurities, resulting in a material with 99% cellulose content and is absorbent and suitably white, also generating huge amounts of effluent (Hossain and Uddin, 2011).

# BIOBLEACHING IN BRIEF

During the pulping process, 90-95% of the lignin is removed by the mills using conventional digesters; the remaining lignin is removed by more selective bleaching agents and the brown unbleached pulp is converted to a white pulp using colored substances (Vidal et al, 1999; Hossain and Uddin, 2011). To reduce the use of chemicals and to retain the pulp strength, several stages of bleaching processes are required. Generally, after each bleaching stage, the pulp is washed to remove any organic material that can be removed with water.

While most of remaining lignin in the unbleached pulp, in the initial stage, the pulp brightened in the concluding stages. For bleaching in the first stage, the pulp's lignin polymer is broken down by oxidants which make the lignin more amenable to reaction with other bleaching chemicals.

In the next stage of bleaching, alkali extraction is performed by exposing the pulp is to a solution of caustic (sodium hydroxide), which degrades most of the lignin so that it dissolves in water and is subsequently washed out of the pulp. In the final stages, for pulp brightening, chlorine dioxide, hydrogen peroxide, or sodium hypochlorite may be used, and generates relatively little organic waste.

The bleached kraft mills in U.S.A, before discharging the effluent from the bleach plant into the mainstream water, undergoes primary and secondary treatment.

Since 1995, all Canadian mills will have secondary treatment systems in order.

All of the bleached kraft mills in Finland have secondary treatment.

Eight of the fifteen Swedish bleached kraft mills have secondary treatment; however, the Swedish mills chose a different strategy to control water pollution. These mills installed oxygen delignification rather than treat the pollution created by the manufacturing process.

To meet the target levels proposed by the Swedish EPA, several mills plan to "close-up" their bleach plants. For mills that lack secondary treatment, the measures to minimize bleach plant effluent flow may be combined with a compact secondary treatment system if necessary to meet the target levels.

## TRADITIONAL BLEACHING: A LOOK AT THE EFFLUENTS

Bleach Plant Effluent Quality and the Manufacturing Technologies We can compare the performance of the manufacturing technologies by examining the loading of biochemical oxygen demand (BOD), chemical oxygen demand (COD), color, adsorbable organic halogens (AOX) and dioxins in the bleach plant filtrates (Pulp and paper manufacturing, chapter 5).

### Biochemical Oxygen Demand (BOD)

BOD is a measure of the tendency of an effluent to consume dissolved oxygen from receiving waters. It is also a measure of the readily biodegradable fraction of pulp mill effluent. Microorganisms in the receiving water consume oxygen as they metabolize the organic material in the effluent. High levels of BOD in the effluent stream can deprive fish, shellfish, fungi and aerobic bacteria of the oxygen they need to survive.

Mills employ secondary treatment systems to remove over 95% of the BOD from the raw effluent. For BOD, environmental impacts are relatively well-controlled by local permitting and monitoring. In most cases, NPDES permits have strict limits based on the assimilative capacity of local receiving waters (Technology-Based Effluent Limitations, 2010). These limits keep BOD discharges below this level to protect aquatic communities.

### *Traditional ECF Bleaching*

Increasing the percentage of chlorine dioxide substitution in the first bleaching stage results in a small decrease in the BOD loading of untreated bleach plant effluent. Secondary treatment systems, designed to remove BOD, are most effective in reducing the BOD of bleach plant effluent and effluent from other sources at the mill. Several experts report that the level of chlorine dioxide substitution has no impact on the magnitude of the BOD loading in bleach plant filtrates when chlorine dioxide substitution in the first bleaching stage increases from 50% to 100%. In fact, there is no consistent evidence that the chlorine dioxide substitution level has any effect on the BOD of bleach plant effluent (Simmons, 1992). These differences in BOD loading at these substitution levels were not statistically different because of the high variability of the data (Environmental footprint comparison tool, 2013; National Council for Air and Stream Improvement, Inc. (NCASI). 2004)

### *Enhanced ECF Bleaching*

Installing extended delignification and/or oxygen delignification decreases the amount of lignin that must be removed in the first bleaching stages, and the BOD loading in the untreated effluent. BOD levels decrease with the decrease in effluent flow. Secondary treatment has the largest impact on the BOD loading in the final bleach plant effluent, although decreasing the amount of BOD going into the treatment plant allows mills to increase production capacity without increasing the size of the effluent treatment system.

## Chemical Oxygen Demand (COD) and Color

Chemical oxygen demand is the amount of oxidizable compounds present in water. The COD of treated mill effluent represents the fraction of the substances in an effluent that the natural ecosystems cannot readily degrade, but provides no indication whether these substances are harmful.85 COD does provide useful information about the sublethal toxicity of ECF and TCF effluents, but the source of COD within the pulp mill provides the most pertinent information. The European Environmental Research Group has performed model ecosystem studies on pulp mill effluents since 1982 that include up to 54 different parameters. They developed a "response index" that summarizes the results of a model ecosystem test on a scale of 1 to 5, where higher numbers correspond to increased effects. The relationship between the response index and COD loading for 14 effluents (Paper Task Force, 1995; Teixeira, 2010) was examined and found that the response index increases with the COD loading in the effluent.

Color measures the large complexes of extracted organic matter from wood. It also is influenced by the presence of metallic ions (e.g., iron and manganese) and humic matter. For kraft pulp and paper mills, most of the effluent color is attributable to losses of pulping liquor (black liquor) and bleach plant extraction stage filtrates. Although color is currently regulated by EPA and states as an aesthetic property, under certain conditions color may interfere with aquatic life by limiting light transmittance. Recent studies of effluent color below bleached kraft mills, however, suggest that it is unlikely that light transmittance is being reduced to a degree that would harm downstream aquatic communities (Vidal et al, 1999). Other studies, based a limited number of kraft mill effluents, have found that the source of much of the color measured in these effluents accounts for the responses in several marine bioassays.

## *Traditional ECF Bleaching Processes*

The impact of different levels of chlorine dioxide substitution on COD, color and AOX for softwood kraft pulp have been studied that has been pulped in a traditional digester (Ragauskas, A. n.d. http://www.ipst.gatech.edu/faculty/ragauskas_art/technical_reviews/Environmental.pdf; National Council for Air and Stream Improvement, Inc., 2004). This pulp was bleached in the laboratory with levels of chlorine dioxide substitution that ranged from 70-100% with different total charges of bleaching chemicals. The COD loading of the bleach plant filtrates ranged from 62 - 73 kg/ADMT while the color loading ranged from 84 -

132 kg/ADMT. While increasing chlorine dioxide substitution from low levels of substitution (< 20%) to high levels (> 50%) decreases the color in bleach plant effluent, "increasing the substitution from 70% to 100% has only a marginal effect on effluent color" (National Council for Air and Stream Improvement, Inc, 2004; Katia et al, 2009)

### *Enhanced ECF Bleaching Processes*

Reducing the amount of lignin in the pulp exposed to chlorine dioxide in the first bleaching stage decreases both the COD and color loading in untreated bleach plant effluent. Hence the role of biobleaching by the use of enzymes or microbes to reduce the quantity of lignin can be of immense value.

Low-effluent ozone ECF and TCF mills recirculate more of their bleach plant filtrate to the chemical recovery system than do mills with enhanced ECF processes. As with BOD loading in untreated bleach plant effluent, COD loading in the untreated bleach plant effluent decreases as the magnitude of the effluent flow decreases. The low-effluent ECF and TCF mills have the lowest COD releases of the six-softwood bleached kraft mills (Maples et al, 1994; Paper Task Force, 1995; Bicknell et al, 1995).

## Adsorbable Organic Halogens (AOX)

AOX is a summary parameter that measures the approximate amount of chlorine (and other halides) present in organic material that adsorbs to activated charcoal; thus, AOX provides an estimate of the total amount of chlorinated organic material in the effluent (Paper Task Force, 1995).

In the regulatory impact assessment of the proposed effluent guidelines, EPA states some of the issues associated with controlling AOX. "Although AOX concentrations can be used to determine the removal of chlorinated organics to assess loading reductions, they do not provide information on the potential toxicity of the effluent, and therefore, are not appropriate to evaluate the potential impacts on the environment. Although no statistical relationship has been established between the level of AOX and specific chlorinated organic compounds, AOX analysis can be an inexpensive method for obtaining the "bulk" measure of the total mass of chlorinated organic compounds."

### *Traditional and Enhanced ECF Bleaching Processes*

AOX loadings in the treated mill effluent decrease as chlorine dioxide substitution in the first bleaching stage increases, and when mills reduce the amount of lignin in the unbleached pulp before it enters the bleach plant.

### *Low-Effluent Bleaching Processes*

The bleach plant AOX loading at the ozone ECF mill is about one order of magnitude less than the AOX loading at the three-chlorine dioxide ECF mills. The bleach plant AOX loading at the two TCF mills is reduced by another one or two orders of magnitude (Paper Task Force, 1995; Pulp and Paper Manufacturing, chapter 5).

## Dioxins

The discussion of dioxin loading in bleached kraft mill effluent has two parts. The first part examines the dioxin loadings in the treated mill effluent for mills that have traditional ECF and enhanced ECF bleaching processes, and the dioxin loadings in the bleach plant effluent for mills with traditional ECF and ozone ECF bleaching sequences. The second part of the discussion explores whether ECF bleaching processes provide an acceptable level of safety from dioxin exposure for certain sensitive populations (United States Environmental Protection Agency, 1997).

### *The current status*

The U.S. industry reacted quickly when EPA found, in 1985, that bleached kraft mills discharged dioxins into their receiving waters. The 1993 NCASI survey of U.S. bleached chemical pulp mills showed that the U.S. industry had reduced the discharge of TCDD 90% from a mean final mill effluent loading of 51 ppq in 1988.

International Paper reported similar results for the final effluent of one of its mills at 50% chlorine dioxide substitution for chlorine in the first bleaching stage (Prakash and Kumar, 2013).

Solomon and his colleagues confirmed the reductions in the concentrations of TCDD and TCDF in the final effluent of mills that increased chlorine dioxide substitution from 50-70% to 100% in an analysis of publicly available data sent to regulatory agencies (Paper

Task Force, 1995; Chee et al, 2016). The TCDD level in all but one of the analyses of final effluent was below a detection limit of 1-10 ppq for mills with traditional ECF bleaching.

### *Ozone ECF and TCF Bleaching Processes*

Much less data currently exists concerning the dioxin concentrations of ECF mills that replace chlorine dioxide with ozone in the first bleaching stage. In their proposed effluent guidelines, EPA requires the concentrations of TCDD and TCDF in the bleach plant filtrates of bleached kraft mills to be below the analytical detection limit of 10 ppq. The concentrations of dioxins will be higher in the bleach plant effluent than in the final effluent because, on average, the bleach plant effluent accounts for about one-third of the total effluent from a bleached kraft mill;108 thus, the dioxin concentration of the untreated bleach plant effluent is a more sensitive measurement of the quantities of dioxins generated during the manufacture of bleached kraft pulp ((Paper Task Force, 1995; Vidal et al, 1999; Katia et al, 2009).

## ENVIRONMENTAL EFFECTS OF BLEACHED KRAFT MILL EFFLUENTS

Several researchers have compared the sublethal toxicity of ECF and TCF bleach plant and mill effluents using bioassays and model ecosystem studies. Sublethal effects include impaired growth, reproduction or immune response, effects that do not kill the organism but may decrease the population of that organism over time.

## STUDIES OF THE SUBLETHAL TOXICITY OF ECF AND TCF BLEACH PLANT FILTRATES

Little public information exists on the environmental effects of moving from the base case to traditional ECF bleaching. Weyerhaeuser128 and Champion International report reduced toxicity from their mill effluent when they converted the bleach plant from the base case to enhanced ECF bleaching with oxygen delignification. It is difficult, however, to attribute these reductions in effluent toxicity directly to the changes in the bleach plant effluent because mills make many additional changes during major modernizations. Most mills upgrade the brownstock washers when they install oxygen delignification or move to complete chlorine dioxide substitution. This improvement, for example, can mask the effect of specific changes in the bleach plant.

The untreated traditional and enhanced ECF bleach plant filtrates were compared, and it was found that filtrates from combined chlorine dioxide and alkali extraction stages from a traditional ECF bleaching sequence reduced water flea reproduction and fathead minnow growth more than did untreated bleach plant filtrates from the oxygen, and alkali extraction stages of an enhanced ECF bleaching process (Paper Task Force, 1995; Norman et al, 1988). While secondary treatment reduced toxicity, both of the enhanced ECF and one of the TCF effluents were still found to be toxic for sensitive sea urchin fertilization tests and scallop larval abnormality tests.

The sensitivity of marine organisms to a component of effluent from bleach kraft mills was also observed. They analyzed different fractions of bleached kraft mill effluent and found that a large, highly-colored water soluble degraded lignin molecule exhibits most of the toxicity to sea urchin sperm, mollusk embryos, sole larvae and kelp sperm and eggs. These researchers found that this "lignin derived macromolecule" acts at the cell surface to inhibit developmental events. This molecule binds to the heads of the sea urchin sperm and thus prevents fertilization of the egg (Paper Task Force, 1995).

Studies suggest that treatment of the substance or bleached kraft mill effluent with reducing agents such as sodium meta-bisulfite effectively reduces the effect on sea urchin sperm. Studies of effluents from mills that use oxygen delignification and extended delignification to produce ECF and TCF pulps suggest that the environmental effects of these processes are low and similar.

The bleach plant effluent from a Finnish mill that switched between enhanced ECF with oxygen delignification and ozone TCF bleaching were compared in a study. Five bioassays were performed: fathead minnow survival and growth, water flea survival/reproduction, Selenastrum (algae) growth, sea urchin fertilization and rainbow trout liver enzyme induction (EROD). In all cases, the treated TCF effluent had the least effect.

While a range of single species bioassay tests show TCF mill effluent to be less toxic than enhanced ECF mill effluent, model ecosystem studies do not show large differences in the environmental effects of enhanced ECF and TCF bleaching of unbleached kraft pulps. Kovacs reported that the model ecosystem studies showed no toxic effects from either effluent at concentrations that ranged from 0.05-0.25%.

## LIFE CYCLE ASSESSMENT

The LCA method emerged in the 1960s and during the 1990s the evolution was rapid. LCA is a method to analyze and assess the environmental impact of a product or service throughout its entire life cycle which includes raw material extraction, processing, transportation, packaging, storage, use and waste management (Straka and Layton, 2010). The flows of energy and material into and out from the system studied are accounted for. The data is aggregated to facilitate interpretation. The contribution to various environmental effects (e.g. climate change) can be assessed.

Life Cycle Assessments are intended to support decisions; the decision makers may be authorities, companies or consumers with respect to products with the same function, different process alternatives or different waste handling alternatives. However, LCA can also be used for individual products with the aim to identify opportunities for environmental improvements. For this, comparisons are made between the different parts of the life cycle. The following are examples of how LCAs have been used by authorities and companies:

- to support legislation and public policy making;
- to improve the environmental performance of products / product development;
- to choose between raw materials and suppliers;
- to make market claims; and
- as input to strategic planning.

## IMPORTANCE OF LIFE CYCLE ASSESSMENT IN BIOBLEACHING:

Thirty years ago, molecular chlorine was commonly used for pulp bleaching.

Currently, **elemental chlorine-free (ECF) bleaching**, using chlorine dioxide ($ClO_2$), is the dominating technology. It has a worldwide share of more than 3/4$^{th}$ of the market for the bleached kraft pulp capacity (Jour et al, 2013).

**Total elemental chlorine-free (TCF) bleaching** does not even dominate even one-tenth of the market share of the bleached kraft pulp.

The published LCAs of pulp and paper products state that the relative significance of the bleaching (i.e., production of bleaching chemicals and the bleaching process at the pulp mill) vary depending on the type of environmental impact (Jour et al, 2013).

## CONVENTIONAL BLEACHING: CASE STUDY AND THEIR IMPACT ON THE ENVIRONMENT:

Two scenarios have been studied in order to investigate the consequences of using different data sets for the bleaching chemicals.

***Chemical Island scenario***: Sodium chlorate produced on-site at the pulp mill, a benefit with this concept is that sodium chlorate can be produced using biomass energy and that less transportation is needed (Jour et al, 2013).

***Ecoinvent scenario***: Carbon footprint data sets for the chemicals chlorine dioxide (with sodium chlorate upstream), hydrogen peroxide and sodium hydroxide (Jour et al, 2013).

## EXPORT OF PULP MILL EXCESS ELECTRICITY

In this study, it is assumed that the mill is connected to the national grid making it possible to export the excess electricity and influences the outcome of the LCA, e.g. carbon footprints of the unbleached pulp.

The carbon footprint can vary between 181 and 202 kg $CO_2$-equivalents per ton of unbleached pulp (Jour et al, 2013). For a pulp mill without a Chemical Island and/or on-site ozone production, 100% of the excess electricity can be exported to the national grid.

A part of the excess electricity is used to produce sodium chlorate when there is a Chemical Island at the pulp mill site, resulting in less excess electricity to export from the pulp mill to the national grid, and a higher carbon footprint, as a consequence. However, this disadvantage is taken care of when looking at the total carbon footprint of the bleached pulp, with an advantage of having access to sodium chlorate/chlorine dioxide with a low carbon footprint, assuming that all of the sodium chlorate produced is consumed on-site in the chlorine dioxide generator (Jour et al, 2013).

When the pulp mill uses ECF ozone bleaching, a part of the excess electricity is needed for the production of ozone. The carbon footprint per ton unbleached pulp is higher in the scenarios involving ozone bleaching. If the scenarios involve both ozone bleaching and a Chemical Island, the carbon footprint will be even higher.

## CARBON FOOTPRINT DATA OF THE BLEACHING AGENTS

There is a large difference in the data for chlorine dioxide, hydrogen peroxide and sodium hydroxide between the Chemical Island and the Ecoinvent scenarios (Jour et al, 2013; Affeldt, 2016). The Ecoinvent data is representative for European manufacturers and assumes an European electricity mix, UCTE, with a quite large share of fossil energy carriers. However, the data found for these chemicals in the Ecoinvent database is much higher. For chlorine dioxide, the main reason for the difference is the electricity mix used in the production of sodium chlorate. The Chemical Island scenario means that the electricity needed for sodium chlorate production is produced from biomass (Jour et al, 2013; Affeldt, 2016).

## CARBON FOOTPRINT DATA OF PULP

The carbon footprint contribution made by the pulp mill (excluding bleaching) is due to production of steam and electricity. To generate energy, not only biomass, but also a small amount of fossil energy is needed. In a Chemical Island scenario it is possible to use hydrogen from the chlorate plant to replace the fossil fuel.

In summary, the carbon footprint of unbleached pulp is roughly made up by

- 43-45% from forestry,
- 44-47% from the pulp mill excluding bleaching,
- 6-7% transport of wood and
- 4% from production of pulping chemicals cradle-to-gate.

The resulting contribution to the carbon footprint by the pulp mill is significant although the share of fossil energy used is low (Jour et al, 2013; Affeldt, 2016). Thus reducing the amount of fossil energy required would improve the carbon footprint of the kraft pulp production.

The sodium chlorate is produced on-site at the pulp mill using biomass for power in the Chemical Island scenario, while in the Ecoinvent scenario, carbon footprint data from the public LCA database Ecoinvent has been used for the chemicals chlorine dioxide (with sodium chlorate upstream), hydrogen peroxide and sodium hydroxide.

## A SUMMING UP OF THE FOOTPRINT

Overall, it can be seen that the carbon footprint contributions made by bleaching are very similar; between 36 and 41 kg CO2-equivalents per ton of bleached pulp for all sequences in the Chemical Island scenario and is much higher in the Ecoinvent scenario, between 97 and 125 kg CO2-equivalents per ton of bleached pulp (Jour et al, 2013).

In all bleaching sequences, sodium hydroxide and chlorine dioxide contribute the most. The chlorine dioxide part also includes the production of sulphuric acid, methanol, sodium chlorate and transportation, if needed, to the ClO2 generator (Gloria et al, 2005). Hydrogen peroxide gives the third largest contribution to the carbon footprint in the ECF light and ECF ozone sequences due to the higher charges used in these sequences. A higher carbon footprint contribution is noticed for the ECF ozone sequence because a higher charge of sulphuric acid is used.

- 15 - 41% of the total carbon footprint of eucalyptus pulp is made up by bleaching including production of bleaching chemicals cradle-to-gate, transportation of bleaching chemicals to the pulp mill and the bleaching process itself.
    - For the Chemical Island scenario, the contribution from bleaching is 15 - 17%.
    - For the Ecoinvent scenario, the contribution from bleaching is 34 - 41%.
- The alternative bleaching sequences studied result in rather similar carbon footprints of the bleached pulp in the Chemical Island scenario. ECF-state of the art shows the lowest carbon footprint.
- There is a large span in carbon footprints of the chemicals used for pulp bleaching. It is crucial to select data that are relevant in terms of geography and technology.

Both sodium chlorate and sodium hydroxide require large amounts of electricity when produced. The way electricity is produced is strongly influencing the carbon footprint of these two chemicals.

- The carbon footprint of unbleached pulp is roughly made up by (Jour et al, 2013):
    - 43-45% from forestry;
    - 44-47% from pulp mill excluding bleaching
    - 6-7% from transport of wood to pulp mill; and
    - 4% from pulping chemicals cradle-to-gate
- assuming that the excess electricity from the mill is exported and replaces electricity produced by the national grid in Brazil.
- A dominance analysis of the relative contribution made by the different greenhouse gases to the total carbon footprint of bleached pulp shows that carbon dioxide gives the most dominant contribution (81-85%), followed by nitrous oxide (~8%), methane (~6%) and non-methane volatile organic compounds (~3%) Jour et al, 2013; Overview of greenhouse gases, 2015).
- For environmental effects other than the carbon footprint, only minor differences between the alternative bleaching sequences have been found.

# BIOBLEACHING

An environmentally friendly alternative to harsh alkaline chemicals is bio bleaching, which uses enzymes to prepare cotton under very mild conditions, generating a waste with less chemicals and a lower volume of water, becoming an economically viable alternative. Enzymes do the same job of preparing the fabric, leaving the cotton fiber, compared to using hot sodium hydroxide to remove the impurities and damaging parts of the fiber. Experts also believe bio bleaching will result in the lowering of BOD, COD, TDS and alkalinity, process time, cotton weight loss and harshness to handle (Csiszár et al, 2001). A detailed study on significance, scope and limitations of biobleaching has been presented in subsequent chapters of this book.

## *A Case Study: Effect Enzyme Bleaching on Effluent Quality and Bio-Treatability, Pulp & Paper*

The bleaching sector of a kraft pulping process is responsible for the high effluent volume generated in modern mills. The effluent is rich in organic matter and contaminants such as organochlorine compounds when chlorine based chemicals are used. The industry has been looking for novel processes in order to minimize the impact of these effluents in the environment. In the 90s, the use of enzymes in pre-bleaching stages aimed at the improvement of effluent quality, particularly the reduction of organochlorine compounds (AOX) in the effluent (Paper Task Force, 1995; Pulp and Paper Manufacturing, chapter 5). Currently, the enzymatic pre-bleaching stages are used mainly for economic reasons. It reduces the consumption of bleaching chemicals, particularly the chlorine dioxide in ECF (elemental chlorine free) sequences. However, when enzymatic pre-bleaching stages are used, an increase of the filtrate's organic load has been observed, compromising the effluent treatment plants (Hossain and Uddin, 2011). Currently, it has been possible to produce stable enzymes by genetic manipulation with high effectiveness for pulp bleaching processes. Industrially, they are commercialized in concentrated liquid suspensions and are applied in small dosages in the pre-bleaching reactor. Two types of enzymes are often used for kraft pulp bleaching: **xylanases** and **laccases** (Woolridge, 2014).

Xylanase has been the most common type of enzyme used for this purpose. It is very efficient and reduces considerably the use of bleaching chemicals (Woolridge, 2014). In some cases, where chlorine or chlorine dioxide is used, it can lower the AOX in the filtrates by as much as 25% (Paper Task Force, 1995; Pulp and Paper Manufacturing, chapter 5). Initial studies demonstrated that the wood type, the bleaching sequence and the type of xylanase could affect the enzymatic treatment. Better conditions of xylanase application have been developed, allowing the use of enzymes in favorable mill conditions, such as alkaline pH, higher temperatures and shorter retention time (Beg et al, 2001). Some **advantages** of xylanase application are

- low capital investment,
- reduction of bleaching chemicals [15-20%],
- low AOX concentration in the final effluent and the higher pulp brightness (Beg et al, 2001; Woolridge, 2014).

However, the effect of xylanases is limited, because

- they do not replace more than 20% of the bleaching chemicals.
- the cost of xylanase is high and
- there is a possibility of pulp yield losses.

Electronic microscopy studies of oxygen bleached pulps followed by enzymatic pre-bleaching had shown some fiber morphologic changes such as holes, cracks, formation of filaments and peeling. The BOD (biochemical oxygen demand) of the filtrate after xylanase pre-bleaching of softwood increases almost two times compared to a reference. Similarly, COD (chemical oxygen demand) and TOC (total organic carbon) also increases. On the other hand, BOD/COD ratio was significantly higher for the xylanase pre-bleaching filtrates, i.e., the effluents seemed to be more biodegradable. In fact, there are only a few studies about the influence of enzyme pre-bleaching on effluent quality (Beg et al, 2001; Jour et al, 2013; Woolridge, 2014).

The enzymatic pre-bleaching has a significant effect on pulp quality. It decreases kappa number and the hexenuronic content of the pulp and increases pulp brightness. However, it also increases pulp yield loss. A substantial increase in the organic matter content (COD) in the filtrates is also observed in the enzymatic pre-bleaching stage. The COD concentration in the filtrates higher than the reference when applied at specific enzyme dosages. In spite of the increase in COD load of the enzymatic pre-bleaching stage filtrates, the generated effluents were efficiently treated in aerobic bioreactors. The COD removal was always superior to, or similar to the reference (Jour et al, 2013; Woolridge, 2014). The increase in the organic matter content in the filtrates will lead to a higher aeration and energy demand in the treatment plant. Moreover, the final COD concentration of the treated effluents from enzymatic pre-bleaching stages will be higher than conventional bleaching sequences.

## SUMMARY

The principles, systems and chemicals of using traditional bleaching systems have been studied in this chapter with special emphasis on their impact on the environment. The carbon foot print along with the gases generated by bleaching industries has been looked at. It has been discussed that several species are affected by the various chemicals present in the effluents associated with traditional pulp bleaching systems. The use of biological agents such as microbes and enzymes can be of significant value in the bleaching of such dyes for two main reasons:

- Their economic cost.
- The use of such natural agents can aid in the reduction of pollution associated with the use of traditional bleaching agents.

## REFERENCES

1. Affeldt C, Leung A, Yang K (2016) Life Cycle Assessment of Stone Paper, Polypropylene Film, and Coated Paper for Use as Product Labels. Master of Science (Natural Resources and Environment) thesis submitted at the University of Michigan.
2. Bicknell B, Spengel D, Holdworth T, "Comparison of Pollutant Loadings from ECF, TCF and Ozone/ Chlorine Di oxide Bleaching," 1995 International NonChlorine Bleaching Conference, p. 16.
3. Chee YT, Pretty MB, Shak KPY, Wu TY (2016) Recent Advancement of Coagulation–Flocculation and Its Application in Wastewater Treatment. Ind Eng Chem Res 55(16):4363–4389
4. Csiszár E, Losonczi A, Szakács G, Rusznák I, Bezúr L, Reicher J (2001) Enzymes and chelating agent in cotton pre-treatment. J

Biotechnol 89:271-279.

5. Woolridge EM (2014) Mixed Enzyme Systems for Delignification of Lignocellulosic Biomass. Catalysts 4:1-35; doi:10.3390/catal4010001
6. Environmental footprint comparison tool. National Council for Air and Stream Improvement, 2013.
7. VidalI G, SotoII M, Méndez RIII, Lema JM (1999) Influence of bleaching technologies on the aerobic biodegradability of effluents from Eucalyptus kraft pulps factories. Braz Arch Biol Technol 42(3).
8. Gloria ZF, Chan A, Minns D (March 2005). The International Journal of Life Cycle Assessment 10(2):136–142.
9. HA Simons Ltd. Assessment of Industry costs to Meet British Columbia's New AOX Regulations. Report prepared for the Ministry of Environment, Lands and Parks and the Ministry of Economic Development, Small Business and Trade of British Columbia, June 1992, III-6.
10. Hossain S, Uddin K (2011) Comparative Analysis Between Conventional Pretreatment and Bio-Preparation. Int J Eng Technol 11(3):11.
11. Jour P, Halldén K, Wackerberg E. Environmental systems analysis of alternative bleaching sequences with focus on carbon footprint. The 46th ABTCP International Pulp and Paper Congress, October, 8-10. 2012, Sao Paulo Brazil. 2013
12. Kátia MME, ColodetteII JL, Silva VL (2009) The role of bound chlorine in the brightness reversion of bleached hardwood kraft pulp. Quím. Nova 32(1) São Paulo. http:// dx.doi.org/10.1590/S0100-40422009000100010
13. Maples G, et al. "BFR: A New Process Toward Bleach Plant Closure," Papers presented at the 1994 International Pulp Bleaching Conference, Vancouver, BC, June 13-16, 1994, 253 - 262. (1994 International Pulp Bleaching Conference hereafter) BFR™ is a licensed trademark of Champion International. Traditional Bleaching and Biobleahing 49
14. National Council for Air and Stream Improvement, Inc. (NCASI). 2004b. Pulp bleaching systems. Chapter 8 in Environmental resource

handbook for pulp and paper mills. Research Triangle Park, NC: National Council for Air and Stream Improvement, Inc.

15. Norman B, Neil M, John BS. The Expert Committee on Kraft Mill Toxicity. Kraft Mill Effluents in Ontario. Municipal Industrial Strategy for Abatement (MISA), March 1988. https://archive.org/stream/kraftmilleffluen00expeuoft/kraftmilleffluen00expeuoft_djvu.txt
16. Overview of Greenhouse Gases. United States Environmental Protection Agency, 2015. https://www.epa.gov/ghgemissions/overview-greenhouse-gases
17. Paper Task Force. White Paper no. 10C Environmental Comparison - manufacturing technologies for virgin and recycled coated paperboard for folding cartons. December 19, 1995.
18. Prakash D, Kumar S (2013) Gas chromatographic studies of chlorophenols on the substitution of chlorine dioxide in Bamboo. Innov Res Chem 1(1):14-21
19. Pulp and paper manufacturing. Chapter 5, https://s3.amazonaws.com/EPNPaperCalc/documents/1685_chapter5.pdf
20. Beg Q, Kapoor M, Mahajan L, Hoondal G (August 2001) Microbial xylanases and their industrial applications: a review. Applied Microbiol Biotechnol 56(3–4):326–338.
21. Ragauskas AND. An introduction – Overview: Environmental aspects, bleaching kraft pulps. IPSTGeorgia Tech Technical Review. http://www.ipst.gatech.edu/faculty/ragauskas_art/technical_reviews/Environmental.pdf
22. Straka TJ, Layton PA (2010) Natural Resources Management: Life Cycle Assessment and Forest Certification and Sustainability Issues. Sustainability 2:604-623; doi:10.3390/su2020604
23. Technology-Based Effluent Limitations. NPDES Permit Writers' Manual, Chapter 5, 2010.
24. Teixeira H (2010) Benthic Macroinvertebrates as Ecological Indicators for Estuarine and Coastal Ecosystems: Assessment and Biobleaching 50 Intercalibration. Doctoral dissertation in Biology (Scientific area of Ecology) presented to the University of Coimbra.

25. United States Environmental Protection Agency (USEPA). EPA eliminates dioxin, reduces air and water pollutants from nation's pulp and paper mills. Press Release November 14, 1997. Washington, DC: United States Environmental Protection Agency.

# CHAPTER 4

# RESEARCH AND FEW CASE STUDIES IN BIOBLEACHING

**Shiv Sanjeevi Sripathi[1] and Prerna Pandey[2]**

[1]Shiv Sanjeevi Sripathi completed his BSc in Biotechnology from Mumbai University and secured second rank in Mumbai University and first rank in his college: Kishinchand chellaram College. Shiv secured first rank in his college KET's V.G.Vaze College and has worked on a stem cell project at the Specialized Centre for Cell Based Therapy (SCCT), KEM Hospital at Mumbai, following which he was a Junior Research Fellow at Microbiology & Cell Biology Department, Indian Institute of Science, Bangalore from August 2009 to November 2011. The field of work involved cloning of cell wall genes and transcription factors in E. coli & M. smegmatis. This was followed by a stint as a Research and Development Executive at Transasia Biomedicals, Mumbai where he worked on patent writing and developing immunoassays from 2011 till 2013. In 2013, he entered the teaching field where he has been tutoring biology and chemistry to the students of 11th and 12th grades along with English grammar and communication skills to people of all age groups.

[2]Dr Prerna Pandey, a biotechnologist with several years of wet lab research experience, worked at International Center for Genetic Engineering and Biotechnology, New Delhi. Her field of research in PhD included isolation and molecular characterization of geminiviruses, genome sequencing, gene annotation, and gene silencing using the RNA interference technology. She has also worked at Transasia Biomedicals and Advance Enzyme Technologies as a scientist. Prerna has published papers in peer-reviewed journals, and has submitted a number of annotated Geminiviral genome sequences in the GenBank, including two novel ones. She has also completed her editing and proof-reading courses from Society for the promotion of Editors and Proof-readers, UK and now works as a freelance scientific editor and writer. When Prerna is not busy with her assignments, she loves to explore historical places.

## INTRODUCTION

In the current world economy, development of environmentally sustainable processes is a challenging task. In this direction, the use of ecofriendly approaches like biocatalysts, enzymes, in various processes is important. For application in various industrial processes, the stability, activity and specificity of enzymes are the required parameters. Identification of newer sources for novel enzymes with desired properties and cost-friendly purification processes is important. Modern approaches such as molecular modeling studies and recombinant DNA technology to develop such novel enzymes are also in great demand.

Individual enzymes are highly specific and work only on a very limited range of substrate, but they function both inside and outside living cells. As proteins, enzymes lose their catalytic properties if heated or subjected to other conditions which denature the protein. There are also biotechnology and environmental benefits.

Man's use of enzyme reactions dates back to antiquity. The making of cheese, vinegar and wine, the leavening of bread, and the brewing of beer are all enzymatic processes which have their origins in prehistory. In the past few decades industries have increased their use of enzymes and enzymes are now widely used in many different sectors. As enzymes can often replace harsh chemicals and help save water, energy and raw materials, they have a positive impact on the environment. What is more, enzymes are completely biodegradable and have no harmful effects in waste water treatment plants or in the aquatic environment at large.

## PEROXIDASE ENZYME FROM VEGETABLES

Plant peroxidases (EC 1.1.11.7) are described as iron containing proteins that oxidize several substrates with the formation of hydrogen peroxide (Chanwun et al, 2013). These enzymes have the potential to be used in several bleaching activities along with the targeted breakdown of the recalcitrant azo dyes and the process of biopulping (Regalado et al, 2004).

These enzymes target several dyes and have found application in the bleaching of several dyes. Perixidases from horseradish, turnip, tomato, soybean, bitter gourd, white radish and *Saccharum uvarum* and several other sources have been examined for their activity (Husain, 2010).

In a study in 2010, peroxidase isolated from fenugreek (*Trigonella foenum-graecum*) showed good biobleaching activity. The authors reported that the enzyme bleached textile effluent process to up to 85% in 5 hours, whereas the complete bleaching of textile effluent by membrane-entrapped FSP was observed within 11 h of its operation (Husain et al, 2010).

A research group examined the potential of a peroxidase enzyme from *Allium sativum*, garlic for bleaching Vat dyes and suggest that garlic peroxidase can find application in biobleaching especially at alkaline pH (Osuji et al, 2014).

## *White rot fungi in the bleaching of effluents*

The role of white rot fungi in the bleaching of effluents had been discussed in Chapter 2. A 1992 study showed the efficiency of *Pycnoporus cinnabarinus*, belonging to the family of white rot fungus. The organism displayed a high efficiency of bleaching of syes when the organism was cultured in a packed bed reactor in a span of 2-3 days (Brahimi-Horn et al, 1992).

## *β-Mannanases*

The β-Mannanases are named as β-1, 4-D-mannan mannohydrolases (EC 3.2.1.78) that can hydrolyze the internal b-1,4-D-mannosidic linkages of mannans. Mannans are considered major portions of the plant cell wall (Tang et al, 2013). These enzymes have been tested for the purpose of bio bleaching, apart from other uses (van Zyl et al, 2010). Certain technical challenges have been reported with the use of these enzymes that include their sensitivity to alterations in pH and temperature. Tang et al, constructed a mannanase gene fusion which had an enhanced activity due to as the addition of a carbohydrate binding group. This recombinant β-mannanase gene was expressed in *P. pastoris* for potential use in industries (Tang et al, 2013).

## *Xylanases*

The plant cell wall is composed of cellulose (35–50%), hemicellulose (of which 20–30% is xylan) and lignin (20–30%). Xylan ranks second in the hemicellulose content and is made up of a linear backbone of b-1, 4-linked xyloses. The hard wood and soft wood content is 15–30% and 7–10%, respectively. It can also be derivatised in the form of heretopolysaccharides to contain groups such as O-acetyl, arabinosyl , 4-O-methyl-D-glucuronic acid ,L-arabinose, D-galactose, D-mannoses, and glucouronic acid through glysosidic bonds with acetic acid and ferulic acid (Walia et al, 2017).

The enzymes called xylanases have been described produced by several organisms in nature that digest xylan present in wood for several purposes. For instance, a 2012 study examined the role of xylanase from *Thielaviopsis basicola* for the purpose of biobleaching of wood pulp (Golugiri et al, 2012). Another 2013 study looked at the potential of a xylanase isolated from Laceyella sacchari for the bleaching of wood pulp (Singh et al, 2013).

The enzyme xylanase has been isolated from several sources such as bacteria present in sea, rumen, other sources on land, apart from marine algae, insects, fungi, several members of crustacean to name a few (Walia et al, 2013). There has been a report of a stable enzyme produced from a Bacullus subtilis species isolated from a marine environment (Annamalai et al, 2009). The enzyme has also been characterized in a water mollusc, *Pomacea insularus* that inhabits fresh waters (Yamaura et al, 1997). Certain reports have detected the enzyme in cucumber seeds and seeds of barley that are germinating (Sizova et al, 2011).

A 2013 paper explored the potential of using xylanase enzyme for the purpose of eliminating the lignin–carbohydrate complex that is generated in the kraft process. The authors looked at an enzyme from Bacillus species as the organism synthesizes xylanases that are extracellular and also show their activity at high pH and a range of temperatures.

The team worked on a novel strain *Bacillus pumilus* SV-85S that was isolated from soil. They saw that the enzyme was resistant to extremes in temperature and was resistant to alkali. They reported a decrease in the usage of chlorine to up to 29.16 % in comparison to other enzymes that had reported a reduction in usage of 20%. The authors also reported that the use of the enzyme aided in the augmentation of tensile strength, breaking length, burst factor, burstness, tear factor, and tearness of the papers tested. This suggested the use of such a system was highlighted by the authors with implications of such a system on minimizing the harmful effects on the environment (Nagar et al, 2013).

A large variety of microbial xylanases are capable of degrading xylan to renewable fuels and chemicals, finding use in food, paper and pulp industries and thus have become a major group of industrial enzymes. Several microorganisms including bacteria, fungi and actinomycetes can readily hydrolyze xylans by producing 1,4-b-D endoxylanases (E.C.3.2.18) and b-xylosidases (EC.3.2.1.37) (Golugiri et al, 2012; Singh et al, 2013).

In recent years, there has been an emphasis on green biotechnology in bleaching process applications for improving the pulp quality and to decrease pollution.

Over the past 15 years, there has been great interest in **biobleaching** and **biopulping** processes (Singh et al, 2013). Enzyme (mainly cellulase-free xylanase) prebleaching technology has been shown to be

- environment friendly and economically cheap;
- it helps decrease the amount of bleaching chemicals required to obtain desired brightness in succeeding chemical bleaching phase.

**Enzyme pre-treatment** also ensures better distribution of NaOH in both hard- and softwoods, and enhances traditional wood-chip pulping and pulp uniformity (Woldesenbet et al, 2012).

Manufactures use huge amount of chemicals in conventional papermaking process, which cause hazardous effluent disposal problems (Verma and Satyanarayana, 2013).

However, the application of xylanase has had a few limitations for its commercial use, the reasons being

- limited hydrolysis of xylans due to their diverged branched nature, thermal instability and narrow pH range, and substrates being less reachable to the enzyme
- end product inhibition and cost of enzyme production (Walia et al, 2013a, b).

Genetic engineering plays an important role in the large-scale expression of xylanase because commercial applications require enzymes at economically viable rates, and thus requires increased production of xylanase at low capital cost.

### *Xylanolytic enzymes*

Structurally, xylan is complex molecule; a linear polymer with various carbon positions substituted with different sugars and/or acidic compounds at the repeating xylopyranosyl groups. For a complex polymer to be efficiently and completely hydrolyzed, requires an array of enzymes with diverse specificity, like:

- Endo-1,4-b D-xylanase (E.C. 3.2.1.8) to randomly cleave the xylan backbone;
- b-D-xylosidases (E.C. 3.2.1.37) to cleave xylose monomers,
- L-arabinofuranosidases (E.C. 3.2.1.55) to help in the removing the side groups iby a-, a-D glucuronidases (E.C. 3.2.1.139) and
- acetylxylan esterases (E.C. 3.1.1.72) to remove acetyl and phenolic side branches (Walia et al, 2017).

Actinomycetes, fungi and bacteria commonly express such a multifunctional xylanolytic enzyme system (Walia et al, 2013a; Driss et al, 2012; Azeri et al, 2010).

### *Extremophilic xylanases*

The optimum activity of xylanases is at neutral (in particular for bacterial xylanase) or slightly acidic (in particular for fungal xylanase) pH and at, or near, mesophilic temperatures (40–60° C) (Walia et al, 2017). However, there are reports of a few produce xylanases are active and stable at extreme pH range of 2 to 11 and temperature range of 5 to 105° C, as well as at high concentration of NaCl-30%, respectively, to adapt to extreme conditions.

## FACTORS AFFECTING THE XYLANASE PRODUCTION

A few organisms produce xylanases constitutively and also through catabolic repression by carbon sources such as glucose or xylose (Walia et al, 2013a, c and 2017). Generally, the best inducers of xylanase have been shown to be

- xylan,
- Small molecular fragments of xylan, like xyloboiose, xyltriose, and cellulose,
- oligosaccharides and their positional isomers of xylose and glucose, which are produced by small amount of constitutively produced enzyme in the medium,
- several synthetic xylosides,
- poor quality paper is an excellent source of carbon and inducer for xylanase in *Thermoascus aurantiacus*, especially in the absence of xylan and xylooligosaccharides.
- cheaper lignocellulosic materials are suitable substrates for solid state fermentation including bran, straw and husk from wheat and rice, sugarcane bagasse, etc.
- certain *Streptomyces* sp and *Bacillus* sp. have shown enhanced production after inclusion of certain amino acids in their media (Beg et al,2000).

# TYPES OF FERMENTATION

Paper pulp industries use xylanase that can be produced at high temperature and pH (Walia et al, 2015). The xylanases from *Thermotoga* sp are the most thermostable and active at 105 ° C for 30 min.

**Both submerged** (SmF) as well as **solid-state fermentation** (SSF) have been used for xylanase production. Bacteria and fungi have been used extensively reported for xylanase production. However, the industrial production of xylanases by **Submerged fermentation** is hindered due to its relatively high cost and being more energy intensive (Walia et al, 2017).

**Solid-state fermentation** (SSF) has more productivity per reactor volume, thus incurring lower operation and capital cost. It is potentially a preferred method for xylanase production, especially for fungal cultivations (Walia et al, 2013, 2017). The factors that affect SSF for enzyme production by microbes include:

- selection of a suitable microorganism and its type and size of the inoculum

- choice of a good substrate, its pre-treatment, and particle size,
- substrate water content and water activity (aw),
- relative humidity,
- temperature control of fermenting and removal of generated metabolic heat,
- time period of cultivation,
- maintaining uniform conditions in SSF system and the gaseous atmosphere, i.e. rates of oxygen consumption during fermentation and evolution of carbon dioxide.

Although SSF is a perfect mode for fungi due to their mycelial nature and ability to grow in less water, it may not be suitable for the growth of all organisms.

## DESIGNS FOR INCREASED PRODUCTION OF XYLANASE

Several experimental designs have been optimized for a number of cultural and nutritional parameters for production of xylanase and other enzymes from microbial and fungal sources (Wang et al, 2008; Guleria et al, 2016a; Walia et al, 2015).

***Response surface methodology*** (RSM) can optimize a fermentation medium for the xylanase production. For instance, there was 3.1- fold increase in production by *Cellulosimicrobium cellulans* CKMX1, by applying RSM The importance of several factors at different levels is realized through applied RSM for the optimization of various factors for the xylanase production in this investigation.

## PURIFICATION AND CHARACTERIZATION OF XYLANASE

For successful application of enzymes in industries, they need to be purified and characterized. Xylanase enzyme can be purified through a series of processes including ammonium sulphate precipitation, gel permeation chromatography, ion exchange chromatography and ultrafiltration (Zhang et al, 2012; Walia et al, 2017). Briefly, crude xylanase preparation is precipitated using ammonium sulfate and substantially concentrating the proteins. From the ammonium sulfate fractional range of 30–80%, the activity could be recovered with maximum at 60–80% (Khandeparkar and Bhosle 2006). In a study, Taibi et al, (2012) have reported that the purified enzyme preparation exhibited nearly 19% of the total activity of the crude form.

Eyzaguirre et al, in 1992 purified one endo-xylanase (molecular mass of 70 kDa) from *Penicillium purpurogenum* using ammonium sulfate fractionation, Gel filtration on Bio-Gel P10, DEAE cellulose and CM Sephadex chromatographies. In a report by Driss et al, (2012), the molecular weight of the xylanase protein was found to be 20–22 kDa calculated through electrophoretic mobility.

The optimum pH for xylanase isolated from different microbes have varied range. Xylanase isolated from *Bacillus* sp. SPS-0 and *Halorhabdus utahensis* have an optimum pH of 6.0 to 8.0 (Waino and Ingvorsen 2003), while some xylanases from other Bacillus strains reached maximum activity at pH 9.0 (Azeri et al, 2010). Some of the other xylanases reported from *Streptomyces cyaneus* SN32, *S. matensis* DW67 and *S. olivaceoviridis* E-86 show stability between pH 4.0 and 9.5 (60° C), 4.5–8.0 (55 ° C) and 2.0–11.0 (30 ° C), respectively (Ninawe et al, 2008; Yan et al, 2009; Kaneko et al, 2000).

Some of the purified xylanases showed optimum activity at temperature around 55° C and the enzyme was stable over the range of 50–60° C, making them useful in pulp biobleaching and some other industrial applications (Walia et al, 2017).

The time-dependent thermal inactivation studies at the optimal temperature of xylanases *Cellulomonas flavigena* Xyl36 and *Cellulomonas flavigena* Xyl53, showed that 60% of the Xyl36 and 50% of the Xyl53 enzyme activity was lost after 1 h at 65 ° C and 55° C, respectively (Santiago-Herna´ndez et al, 2007).

## BENEFITS OF GENETIC ENGINEERING

In the genetic material of all living cells, a single gene contains the code for the production of a particular enzyme or other protein. Through genetic engineering genes from one type of organism can be transferred to another. In this way, the desired amounts of naturally occurring enzymes can be produced in microorganisms that are suitable for large-scale controlled fermentation. The use of genetic engineering in fermentation technology offers a number of clear benefits for both the environment and consumers, such as lower consumption of raw materials, water and energy.

Several enzyme manufacturers use only well-known and safe microorganisms for enzyme production. Genetic engineering is applied only to benefit optimally from production plants and raw materials, which is a clear advantage for the environment. During the production of the enzyme products and downstream processing, the absence of genetically modified organisms (GMO) is absolutely ensured.

The use of synthetic dyes in several industries necessitates the use of a bleaching step to remove these harmful dyes. Though there are several physical and chemical methods to remove these dyes, the use of biological methods serves added merits such as ease and limited effects of the methods used.

A few case studies shall be studied to get a glimpse of several activities that have been carried out.

## ROLE OF RECOMBINANT DNA TECHNOLOGY

Most of the reported xylanases do not meet the demand of industries, because they show low yield and incompatibility and are thus non-economical Walia et al, 2017. Recombinant xylanase genes have been cloned and expressed in heterologous as well as homologous host systems. At the industrial level, xylanase production in heterologous system has been used as the main tool.

Compared to the native enzymes, recombinant xylanases have shown more industrially useful properties. Xylanase from bacteria such as *Cellulosimicrobium* sp., *Cellulosimicrobium cellulans* have been cloned and expressed; *Thermobifida halotolerans*, *Nesterenkoniaxinjiangensis*, and *Bacillus subtilis* into a non-cellulase producing strain of *E. coli* (Walia et al, 2017). The *xylanase* and *xylosidase* genes were introduced through cloning for the improvement of fermentation processes of industrially important xylose and for enhancing of xylanolytic activity devoid of cellulase activity.

*E. coli*, Bacillus sp., Lactobacillus sp., *Saccharomyces cerevisiae* and filamentous fungi are non-toxic and generally recognized as safe

(GRAS) and have been attractive hosts for research and for production of heterologous proteins (Juturu and Wu, 2011). The ability of performing eukaryotic posttranslational modifications makes **yeast systems** more attractive for heterologous protein expression. Yeast can also grow to very high cell densities and secrete proteins into the fermentation media, thus eliminating and additional step of cell lysis for enzyme purification (Ahmed et al, 2009). One of the best examples is the yeast *Pichia pastoris* has an efficient system for heterologous proteins expression under the control of its own specific promoters.

***Filamentous fungi***, via both homologous and heterologous gene expression, are capable producers of xylanases and influence high expression yields (Ahmed et al, 2009). For use in food industry, the *xylanase II* gene from *Aspergillus usamii* was cloned into the vector pPIC9 K by electroporation technique and integrated into the genome of *P. pastoris* KM71. The enzyme XynII was produced extracellularly in *P. pastoris* following methanol induction (Zhou et al, 2009).

Driss et al, (2012), constitutively expressed high levels of xylanase in *P. pastoris* X-33 by using xylanase GH11 from *Penicillium occitanis* Pol6. The recombinant xylanase expressed extracellularlyunder the control of constitutive promoter glyceraldehyde 3-phosphate dehydrogenase (GAP.

## APPLICATION OF XYLANASE IN PULP AND PAPER MAKING

The pulp and paper industry has been continuously searching for alternate, novel biotechnological methods to replace the chemical processes for making paper (Verma and Satyanarayana, 2013).

***Biobleaching*** is the enzymatic removal of lignin in the hemicellulosic fraction with little harm to the pulp. Xylanases help in bleaching for removal of lignin, expanding pulp fibrillation, decreasing beating times in specific pulp and expanding flexibility in fibers that have to be reused.

Prebleaching using xylanase is a cheaper as well as an environment friendly innovation. It can significantly reduce the amount of bleaching chemicals required to achieve a given brightness. The effectiveness of chemical extraction of lignin from pulp is enhanced through pretreatment with xylanases and minimizes the necessity of chlorine dioxide (ClO2) (Golugiri et al, 2012; Walia et al, 2017).

***Biopulping*** involves pretreating wood or non-wood by lignin-degrading fungi before the routine pulping process. The time required in the pretreatment is around 2–4 weeks.

In biopulping, prior to mechanical or chemical pulping the wood chips and other lignocellulosic materials with natural wood decay are treated with fungi. According to mill operations, the wood is then debarked, chipped and screened. Then chips are briefly steamed to reduce the natural microbial load present in wood chips, and are cooled with forced air and inoculated with the fungus that carries out biopulping. The main advantages of biopulping are:

- reduced consumption of electrical energy
- better throughput of the mill for mechanical pulping, and
- process also improves strength properties of paper

# BIOPULPING FOR THE KRAFT PROCESS

The use of biopulping is still under research for pretreatment for the Kraft process. Moreover, the uses biopulping for non-woody plants like straw, corn stalks and other substrates also needs to be studied (Woldesenbet et al, 2012).

In the conventional pulping process, cellulose degradation is of main concern, which invariably affects the cellulose fiber and subsequently the paper quality (Walia et al, 2017). The use of chemical cannot fully remove the lignin from the fiber some parts of lignin and its derivatives get reprecipitated on to the fiber surfaces, giving the resultant pulp gets a characteristic brown color.

Enzymatic treatments of pulp using xylanase is both economical and gives better fiber qualities. Due to their not too tight structures, xylans are more accessible to hydrolytic enzymes, and give specific activity 2–3 times more than the other polymers hydrolases like that of crystalline cellulose (Azeri et al, 2010; Shatalov and Pereira 2008).

The xylanase treatment before chemical bleaching may be effective due to cleavage of residual lignin linked to hemicellulose. This increases the pulp accessibility to bleaching chemicals, thereby enhancing lignin extraction during subsequent bleaching stages (Walia et al, 2017). Briefly, biobleaching has several advantages:

- decreased use of bleaching chemicals, like organic halogen compounds,
- improved pulp and paper quality, and
- reduced release of toxic effluents.

The entrapped lignin from the fiber wall may be diffused due to the carbohydrate complexes from the released xylan. Limited removal of pulp xylan helps to increase the pulp bleachability during subsequent bleaching stages (Walia et al, 2017). In the presence of cellulose, slight decrease in inter-fiber bonding strength is seen in enzyme treated sheets without affecting the fiber's mechanical strength (Valls et al, 2010). Xylanase increases viscosity in the absence of cellulose (Li et al, 2010).

***Commercial xylanases***: crude xylanase from *C. cellulans* CKMX1 showed high thermostability (up to 60° C) over a pH range of 5–10 and brought reduction of highest kappa number from 0.5 and 0.8 points with gain in brightness of 0.93 and 1.42% ISO points, respectively. This implies savings of up to 1/8$^{th}$ of the total chlorine consumption with straw pulp of reasonable quality (Walia et al, 2017). Xylanase from *A. niger* was shown to bring down total chlorine consumption by 1/5$^{th}$, with an acceptable increase in brightness, respectively (Walia et al, 2017). The essential characteristics of xylanases for their usefulness in pulp and paper industry are:

- samples free from cellulase after treatment,
- they are stability in alkaline conditions, and
- thermostable,
- because of their low molecular weight, they can easily penetrate into the xylan on the surface of pulp particles (Valls et al, 2010).

# PEROXIDASES

## *Application of peroxidases in de-colorization of industrial dyes*

Enzymes such as manganese peroxidases (MnPs), and lignin peroxidases (LiPs) are used for the decolorization of synthetic azo dyes such as orange II, and others (Pandey et al,2017). Horseradish peroxidase (HRP) can degrade phenol and substituted phenols and is effective in degrading and precipitating Remazol and other azo dyes that are industrially important (Bhunia et al, 2001). The white-rot fungus expresses the enzymes LiP, MnP, and laccases, and is able to oxidize several recalcitrant xenobiotics released to the environment (Pandey et al, 2017).

The textile industry consumes large quantities of water, and chemicals like synthetic dyes and releases large volumes of wastewater. Among all industrial wastes, the wastewater from textile industries is one of the most polluting, and requires appropriate treatment strategies.

**Peroxidases** have been significantly effective in decreasing textile industry pollutant residues (Husain, 2006; Maciel et al, 2007). Peroxidases, such as horseradish, Brassica campestris turnip, bitter gourd, tomato, soybean, Ipomea palmata and *Saccharum spontaneum* peroxidases, are being used for degrading and detoxifying polychlorinated biphenyls, polyaromatic hydrocarbons, and other synthetic industrial dyes (Baldrian and Snajdr, 2006).

Application of peroxidases in pulp and paper industry

Lignin, a phenolic heteropolymer, hinders a proper exploitation of plant biomass for the pulp and paper industry and should be removed for production of good quality paper. The paper industry has been successfully using **Lignin peroxidase** (LiP) and **manganese peroxidase** (MnP) for processes like biopulping, biobleaching, and selective delignification (Hatakka et al, 2003; Aehle, 2007).

# LACCASE

Until recently, although the substrate range of laccases was believed be restricted to phenolic subunits, they have long been thought to play some role in biodegradation of lignin (Higuchi, 1989). Laccases are produced abundantly by *T. uersicolor* during pulp bleaching and at least three forms can be isolated. Each form appears to bind differently to pulp. In the presence of certain easily oxidized substrates, such as 2,2'- azinobis(3-ethylbenzthiazoline-6-sulfonate) (ABTS), all three laccases could oxidize non-phenolic lignin substructures (Reid and Paice, 1994). Moreover, the combination of ABTS and laccase was shown to delignify kraft pulp, and a decrease in both kappa number and Klason lignin content was observed (Reid and Paice, 1994).

## MANGANESE PEROXIDASE

Initially, manganese and lignin peroxidase were not detected in bleaching cultures of *Trametes versicolor* (Archibald, 1992). It was through Mn(III) malonate formation [16], a more sensitive assay for manganese peroxidase, the enzyme could be detected and isolated from such cultures. The appearance of manganese peroxidase during bleaching was somewhat unexpected, because of a previous report (Reid and Paice, 1994) that stated that secretion of manganese peroxidase by *T. versicolor* occurred in unagitated, nutrient-limited cultures only. The activity of extracellular manganese peroxidase and maximum rate of pulp bleaching peaked at the same time (Paice et al, 1993). The isolated manganese peroxidase generated methanol from pulp and decreased the lignin content (kappa number) in the presence of hydrogen peroxide and malonate, although the brightness of the pulp was not significantly increased. In addition, manganese peroxidase that were partially purified alone could demethylate and delignify pulp nearly to the same extent as the complete culture supernatant from bleaching cultures. The pulp induced the enzyme, probably because of the manganese content of such pulps (Paice et al, 1993), which is available for the enzyme's catalytic cycle. In the presence of the pulp, *T. versicolor* also produces carboxylic acid chelators such as oxalate (Roy and Archibald, 1993; Reid and Paice, 1994).

## SUMMARY

Our society, especially the scientific world has become more aware of the impact textile mills have on the environmental. The textile effluent quality can be improved by using enzymes. It is obvious that enzymatic process need less number of chemicals, and subsequently less amount of water, energy & time. Xylanase enzyme as a pre-bleaching agent has been established in various laboratories and also been commercially used in several textile industries. Although several commercially available enzymes are available, it has important to develop indigenous enzymatic pre-bleaching process from locally isolated cultures because of the differences in paper making process in the developed counties and India. Search for suitable xylanases from various sources is a continuous process. Another industrially important enzyme are the peroxidases that catalyze oxidation of a wide range of phenolic as well as non-phenolic compounds, and in bioremediation of environmental pollutants.

## REFERENCES


1. Aehle W (2007) Enzymes in industry: Production and Applications. Wiley, New York. ISBN-13: 978-3527316892.
2. Ahmed S, Riaz S, Jamil A. (2009). Molecular cloning of fungal xylanases: an overview. Appl Microbiol Biotechnol. 84:19–35.
3. Annamalai N, Thavasi R, Jayalakshmi S, Balasubramanian T. (2009). Thermostable and alkaline tolerant xylanase production by Bacillus subtilis isolated from marine environment. Ind J Biotechnol 8:291–297.
4. Archibald FS. (1992) The role of fungus-fiber contact in the biobleaching of kraft brownstock by Trametes (Coriolus) uersicolor. Holzforschung 46, 31)5-310
5. Azeri C, Tamer AU, Oskay M. (2010). Thermoactive cellulase-free xylanase production from alkaliphilic *Bacillus* strains using various agro-residues and their potential in biobleaching of kraft pulp. Afr J Biotechnol. 9(1):63–72
6. Baldrian P, Snajdr J. (2006) Production of ligninolytic enzymes by litter-decomposing fungi and their ability to decolorize synthetic dyes. Enzyme Microb Technol 395: 1023-1029.
7. Bhunia A, Durani S, Wangikar PP. (2001) Horseradish peroxidase catalyzed degradation of industrially important dyes. Biotechnol Bioeng 72: 562-567.
8. Brahimi-Horn MC, Lim K, Liany SL. (1992) Binding of textile azo dyes by Myrothecium verrucaria. J Ind Microbio 10:31-36.
9. Chanwun T, Muhamad N, Chirapongsatonkul N, Churngchow N. (2013). *Hevea brasiliensis* cell suspension peroxidase: purification, characterization and application for dye decolorization. AMB Express 3, article 14.
10. Driss D, Bhiri F, Elleuch L, Bouly N, Stals I, Miled N, et al, (2012). Purification and properties of an extracellular acidophilic endo-1, 4-b-Xylanase, naturally deleted in the "thumb", from *Penicillium occitanis* Pol6. Proc Biochem 46:1299–1306.

11. Eyzaguirre J, Scarpa J, Belancic A, Steiner J. (1992). The xylanase system of *Penicillium purpurogenum*. In: Visser J, Beldman G, Kusters-van Someren MA, Voragen AGJ, editors. Xylans and xylanases. Amsterdam: Elsevier; 1992: 505–510
12. Golugiri BR, Thulluri C, Cherupally M, Nidadavolu N, Achuthananda D, Mangamuri LN, et al, (2012) Potential of thermo and alkali stable xylanases from Thielaviopsis basicola (MTCC-1467) in biobleaching of wood Kraft pulp. Appl Biochem Biotechnol 167:2369–2380.
13. Guleria S, Walia A, Chauhan A, Shirkot CK. (2016). Optimization of milk clotting enzyme production by *Bacillus amyloliquefaciens* SP1 isolated from apple rhizosphere. Bioresour Bioprocess.
14. Hatakka A, Lundell T, Hofrichter M, Maijala P. (2003). Manganese peroxidase and its role in the degradation of wood lignin. In: Applications of enzymes to lignocellulosics. Mansfield SD, Saddler JN (Edn) ACS Symposium Series 855, Washington, 230-243.
15. Higuchi T. (1989) Mechanisms of lignin degradation by lignin peroxidase and laccase of white-rot fungi. In: Plant Cell Wall Polymers (Lewis, N.G, and Paice, M.G. Eds.), ACS Symposium Series Vol. 399, pp. 482-502. American Chemical Society, Washington, DC.
16. Husain Q, Karim Z, Banday ZZ. (2010). Decolorization of textile effluent by soluble fenugreek (*Trigonella foenum-graecum* L) seeds peroxidase. Water Air Soil Pollut. 212(1–4): 319–328.
17. Husain Q. (2006). Potential applications of the oxidoreductive enzymes in the decolorization and detoxification of textile and other synthetic dyes from polluted water: a review. Crit Rev Biotechnol 60: 201-221.
18. Husain Q. (2010). Peroxidase mediated decolorization and remediation of wastewater containing industrial dyes: a review. Rev Environ Sci Biotechnol 9(2): 117–140.
19. Juturu V, Wu JC. (2014). Microbial exo-xylanases: a mini review. Appl Biochem Biotechnol. 174:81–92. doi: 10.1007/s12010-014-1042-8.
20. Kaneko S, Kuno A, Muramatsu M, Iwamatsu S, Kusakabe I, Hayashi K. (2000). Purification and characterization of a family

G/11 β-xylanase from *Streptomyces olivaceoviridis* E-86. Biosci Biotechnol Biochem. 64:447–451.

21. Khandeparkar R, Bhosle N.(2006). Purification and characterization of thermoalkalophilic xylanase isolated from the *Enterobacter* sp. MTCC 5112. Res Microbiol 157:315–325.
22. Li X, She Y, Sun B, Song H, Zhu Y, Lv Y, Song H. (2010). Purification and characterization of a cellulase-free, thermostable xylanase from *Streptomyces rameus* L2001 and its biobleaching effect on wheat straw pulp. Biochem Eng J 52(1):71–78.
23. Maciel HPFG, Gouvea CMCP, Pastore GM. (2007). Extraction and partial characterization of peroxidase from *Copaifera langsdorffii*Desf. leaves. Cienciae Tecnologia de Alimentos 27: 221-225.
24. Nagar S, Jain RK, Thakur VV, Gupta VK. (2013). Biobleaching application of cellulase poor and alkali stable xylanase from Bacillus pumilus SV-85S. 3 Biotech 3:277–285.
25. Osuji AC, Eze SOO, Osayi EE, Chilaka FC. (2014). Biobleaching of Industrial Important Dyes with Peroxidase. Partially Purified from Garlic. Hindawi Publishing Corporation. Scientific World Journal. 2014, Article ID 183163, 8 pages. http://dx.doi.org/10.1155/2014/183163.
26. Paice MG, Jurasek L, Ho C, Bourbonnais R, Archibald F. (1989). Direct biological bleaching of hardwood kraft pulp with the fungus Coriolus t'ersicolor. Tappi J. 72(5), 217-221.
27. Paice MG, Reid ID, Bourbonnais R, Archibald FS, Jurasek L. (1993). Manganese peroxidase, produced by Trametes t:ersicolor during pulp bleaching, demethylates and delignifies kraft pulp. Appl. Environ. Microbiol. 59, 260-265.
28. Pandey VP, Awasthi M, Singh S, Tiwari S, Dwivedi UN. (2017). A Comprehensive Review on Function and Application of Plant Peroxidases. Biochem Anal Biochem 6:308.
29. Regalado C, Garc´ıa-Almend´arez, B, Duarte-V´azquez MA. (2004). Biotechnological applications of peroxidases. Phytochemistry Reviews 3(1-2): 243–256.

30. Reid ID, Paice MG. (1994). Biological bleaching of kraft pulps by white-rot fungi and their enzymes. FEMS Microbiology Reviews 13 (1994) 36q-376
31. Roy BP, Archibald FS. (1993). Effects of kraft pulp and lignin on Trametes versicolor carbon metabolism. Appl Environ Microbiol. 59, 1855-1863.
32. Santiago-Hernández A, Vega-Estrada J, Montes-Horcasitas MC, Hidalgo-Lara ME. (2007). Purification and characterization of two sugarcane bagasse-absorbable thermophilic xylanases from the mesophilic *Cellulomonas xavigena*. J Ind Microbiol Biotechnol 34:331–338.
33. Shatalov AA, Pereira H. (2008). Effect of xylanases on peroxide bleachability of eucalypt (*E. globulus*) kraft pulp. Biochem Eng J. 40:19–26.
34. Singh V, Pandey VC, Agrawal S. (2013). Potential of Laceyella sacchari strain B42 crude xylanase in biobleaching of kraft pulp. Afr J Biotechnol 12(6):570–579.
35. Singh V, Pandey VC, Agrawal S. (2013). Potential of *Laceyella sacchari* strain B42 crude xylanase in biobleaching of kraft pulp. Afr J Biotechnol 12(6):570–579.
36. Sizova MV, Izquierdo JA, Panikov NS, Lynd LR (2011). Cellulose and xylan-degrading thermophilic anaerobic bacteria from biocompost. Appl Environ Microbiol 77(7):2282–2291.
37. Taibi Z, Saoudi B, Boudelaa M, Trigui H, Belghith H, Gargouri A, et al, (2012). Purification and biochemical characterization of a highly thermostable xylanase from *Actinomadura* sp. strain Cpt20 isolated from poultry compost. Appl Biochem Biotechnol 166(3):663–679.
38. Tang C-D, Li J-F, Wei X-H, Min R, Gao S-J, et al, (2013). Fusing a Carbohydrate-Binding Module into the Aspergillus usamii b-Mannanase to Improve Its Thermostability and Cellulose-Binding Capacity by In Silico Design. PLoS ONE 8(5): e64766. doi:10.1371/journal.pone.0064766.
39. Valls C, Vidal T, Roncero MB. (2010). The role of xylanases and laccases on hexenuronic acid and lignin removal. Proc Biochem 45:425–430.

40. van Zyl WH, Rose SH, Trollope K, Go¨rgens JF. (2010). Fungal b-mannanases: Mannan hydrolysis, heterologous production and biotechnological applications. Proc Biochem 45: 1203–1213.

41. Verma D, Satyanarayana T. (2013). Production of cellulase-free xylanase by the recombinant *Bacillus subtilis* and its applicability in paper pulp bleaching. Biotechnol Prog.

42. Waino M, Ingvorsen K.(2003). Production of β- xylanase and β-xylosidase by the extremely halophilic archaeon *Halorhabdus utahensis*. Extremophiles. 7:87–93.

43. Walia A, Guleria S, Mehta P, Chauhan A, Parkash J. (2017). Microbial xylanases and their industrial application in pulp and paper biobleaching: a review. Biotech 7:11.

44. Walia A, Mehta P, Chauhan A, Shirkot CK. (2013). Optimization of cellulase-free xylanase production by alkalophilic Cellulosimicrobium sp. CKMX1 in solid-state fermentation of apple pomace using central composite design and response surface methodology. Ann Microbiol 63:187–198.

45. Walia A, Mehta P, Guleria S, Shirkot CK. (2015). Improvement for enhanced xylanase production by *Cellulosimicrobium cellulans* CKMX1 using central composite design of response surface methodology and its application in biobleaching. Biotech 5:1053–1066.

46. Wang Q, Hou Y, Xu Z, Miao J, Li G. (2008). Optimization of cold active protease production by the psychrophilic bacterium *Colwellia* spp. NJ341 with response surface methodology. Bioresour Technol 99:1926–1931.

47. Woldesenbet F, Virk AP, Gupta N, Sharma P. (2012). Effect of microwave irradiation on xylanase production from wheat bran and biobleaching of eucalyptus Kraft pulp. Appl Biochem Biotechnol. 167:100–108.

48. Yamaura IT, Koga T, Matsumoto KV. (1997). Purification and some properties of endo-1,4-b-D-xylanase from a fresh-water mollusc, *Pomacea insularus* (de Ordigny). Biosci Biotechnol Biochem. 61:615–620.

49. Zhang F, Hu SN, Chen JJ, Lin LB, Wei YL, Tang SK, et al, (2012). Purification and partial characterisation of a thermo-stable xylanase from salt-tolerant *Thermobifida halotolerans* YIM90462T. Proc Biochem 47:225–228.

50. Zhou C, Wang Y, Wu M, Wang W, Li D. (2009). Expression of Xylanase II from *A. usamii* in *P. pastoris.* Food Technol Biotechnol. 47(1):90–95.

# CHAPTER 5

# BIOBLEACHING: STATUS, ROADBLOCKS AND FUTURE

**Shiv Sanjeevi Sripathi[1] and Prerna Pandey[2]**

[1]Shiv Sanjeevi Sripathi completed his BSc in Biotechnology from Mumbai University and secured second rank in Mumbai University and first rank in his college: Kishinchand chellaram College. Shiv secured first rank in his college KET's V.G.Vaze College and has worked on a stem cell project at the Specialized Centre for Cell Based Therapy (SCCT), KEM Hospital at Mumbai, following which he was a Junior Research Fellow at Microbiology & Cell Biology Department, Indian Institute of Science, Bangalore from August 2009 to November 2011. The field of work involved cloning of cell wall genes and transcription factors in E. coli & M. smegmatis. This was followed by a stint as a Research and Development Executive at Transasia Biomedicals, Mumbai where he worked on patent writing and developing immunoassays from 2011 till 2013. In 2013, he entered the teaching field where he has been tutoring biology and chemistry to the students of 11th and 12th grades along with English grammar and communication skills to people of all age groups.

[2]Dr Prerna Pandey, a biotechnologist with several years of wet lab research experience, worked at International Center for Genetic Engineering and Biotechnology, New Delhi. Her field of research in PhD included isolation and molecular characterization of geminiviruses, genome sequencing, gene annotation, and gene silencing using the RNA interference technology. She has also worked at Transasia Biomedicals and Advance Enzyme Technologies as a scientist. Prerna has published papers in peer-reviewed journals, and has submitted a number of annotated Geminiviral genome sequences in the GenBank, including two novel ones. She has also completed her editing and proof-reading courses from Society for the promotion of Editors and Proof-readers, UK and now works as a freelance scientific editor and writer. When Prerna is not busy with her assignments, she loves to explore historical places.

# INTRODUCTION

The key impact on the pulp and paper industry was growing consumer demand for environmentally safe products and processes. It put forth its own sway on the environment, like any other large-scale industry. Pulp mills still relief unpleasantly sensing sulfur compounds into the air and discharge waste waters which enrich eutrophication and have deadly effects on biota in the receiver. Recently perhaps most attention attracted by organic chlorine compounds, arise mainly from the reactions between residual lignin present in wood fibers, bring about the brown color of unbleached pulp, and the chlorine used during the bleaching of chemical. For the same amount of active chlorine used, chlorine dioxide causes the formation of AOX in an amount of only one-fifth of that caused by chlorine gas. Earlier measures taken by the forest industry to solve the chlorine problem have focused on measures external to the process but now the emphasis has shifted more towards improving the processes (Kristina and Ann-Kristin, 2013). Between the bleaching technologies, enzymes have already shown their potential, can be used for specific reforms of pulp for the development of environmentally safe processes.

The basic idea of enzyme-aided bleaching was first published in 1986 at the VTT Biotechnical Laboratory in co-operation with the Finnish Pulp and Paper Research Institute aiming to explore the use of enzymes capable of degrading either lignin or hemicellulose as a means of developing a pulp bleaching process involving no chlorine (Bajpai, 2004). During major research interest, the idea of using a hemicellulase treatment to promote pulp bleaching arose but, due to lack of clarity of the enzymatic degradation mechanisms of the residual macromolecular lignin, their practical use has curved to be extremely difficult. Originally idea was to hydrolyze a relatively low number of hemicelluloses in pulps.

During the early phases, different types of hemicellulases were produced by fungal and bacterial strains and the enzymes were compared in bleaching tests. The improvement in bleachability was evaluated by using peroxide delignification or conventional chlorine bleaching (Bajpai, 2004). It was clear that the brightness was largely increased by the enzymatic usage could be exploited for reduction of chlorine chemical consumption. It was observed that the saving could be grown either in the prebleaching stage, as chlorine gas or in the final bleaching stage, as chlorine dioxide. The possibility of minimizing the consumption of chlorine gas in bleaching was of particular interest for environmental reasons. the utilization of enzymes as alternative bleaching agents has become realistic as enzymatic treatment was combined with more efficient pulping and non-chlorine bleaching methods.

## MITIGATION

Some of the result of the pulp and paper industry can be discussed and sustainable practices can be changed. Day by day, old growth forests are getting lost, only due to excessive use of wood from plantation; they are apparently used for the bleaching process to produce TCF paper. Additionally, the emission of carcinogenic organochlorines can be significantly reduced with the help of non-elemental chlorine like peracetic acid, ozone and $H_2O_2$.

## PAPER MILLS AND PAPER RECYCLING

According to the United States Environmental Protection Agency (USEPA), production of virgin paper needs more exploitation of resources, such as water and pollution; contrary to which, the recycled paper can be bleached, without using any extra chemicals. If no chlorine compounds are used in the recycling process, the recycled pulp is identified as the PCF, i.e. process chlorine free.

Thus, recycling the paper can be acknowledged as one of the least complicated procedures and the most feasible alternative to the current paper industry, which will be helpful in minimizing the cost and the environmental pollution.

A **deinking** mill serves as a step in the recycling paper process; it detaches the ink from the paper fibers, along with any other excess materials which are also removed from the remaining paper, it is more time consuming costly than the well-developed paper mills that create paper with the use of trees. But, in environmental perspective it has a multitude of benefits (Palmer, 1995; Environmental Impact of Paper).

## REGULATIONS

Pulp and paper industry is identified to be one of the highly controlled and regulated industry in North America; the US and the Canadian regulations regarding the same have been summarized in this chapter.

## AIR AND WATER POLLUTION

In 1974, EPA was the first organization to issue national waste water standards, as per the Clean Water Act. The regulation has identified numerous effluent limitations for several conventional pollutants. (Gray and Shadbegian, 2015; EPA, 1982)

The intent of this regulatory decision was to improve more industry clarity towards the compliance and achieving higher level of pollution prevention. It should as well be noted that these standards are incorporated by National Pollutant Discharge Elimination System (NPDES) permits, which are to be renewed every five years.

The regulations related to the air can be applicable to the mills that are involved in applications, such as chemical pulping, condensation of hazardous chemicals and reduction of volatile organic carbon as well as other particulate matter. Additionally, these mills need to be compliant with waste water regulations as well, since they are involved in the activities such as chlorine bleaching with kraft chemical pulping with an intention to reduce dioxins, furan and chloroform discharge by almost 96% or more than that. (Gray and Shadbegian, 2015)

Since, the EPA regulations wanted industry to incorporate the best management practices, related to the process control monitoring (EPA, 1998); it promoted some add on numeric effluent limitations for 12 chlorinated phenolic pollutants and absorbable organic halides (AOX). The EPA standards revised the NESHAP regulations in the year 2012, especially for pulp and paper industry; according to which, standards have been set for pollutants, like carbon monoxide, nitrogen oxide, ozone, sulfur dioxide and other particulate matter. (EPA, 2016)

## GLOBAL AIR AND WATER POLLUTION GUIDELINES

According to the Environmental, Health and Safety (EHS) guidelines, federal regulations have developed factories related to many industries and companies, which are compliant to limit air and water pollution to the significant levels.

## MECHANICAL PULP MILLS AND BLEACHING MECHANICAL PULPS

The process of production of wood pulp, just by mechanically grinding wood is known to be mechanical pulp, which is mainly used in the newsprint. The mechanical pulp is identified to be a suitable alternative to kraft or sulphite mills, since it requires fewer chemicals and chemicals that are less toxic. For example, the resin acids released from the wood can be a primary source of pollution, but its effect can be minimized with the help of chemicals that are less toxic.

## INKS

The use of volatile organic compounds, heavy metals and non-renewable oils are main issues with the environmental impact of printing inks, for which standards have been set (Environmental impact of printing inks, 2013). By the year 1970, the US federal regulations, emphasized on the minimum use of toxic metals, such as lead, arsenic, mercury, selenium, cadmium, and hexavalent chromium.

## AIR POLLUTION

One should understand that the pulp and the paper industry would have been a great reason for air pollution, since it involves active emission of hazardous chemicals, like lead, hexachlorobenzene, dioxins, furans and polycyclic aromatic hydrocarbons. Equally, in Canada, less than 2% of total emissions in 2015 were of these chemicals (Environment and Climate Change Canada. 2016).

Other than some air releases mentioned above, some important pollutants include particulate matter of 2.5 microns diameter bits or less, which can offer a serious health hazard by entering into the respiratory system of many young individuals. The industry in the U.S. and Canada is responsible for roughly 10% of the industrial releases of $PM_{2.5}$ (Environment and Climate Change Canada. 2016).

## WATER POLLUTION

Waste water discharge from pulp and the paper mills is comprised of solids, nutrients and soluble organic matter, like lignin, alcohols, chelating agents and inorganic material, such as chlorates as well as transition metals compounds. Their releases have worsened eutrophication of fresh water bodies, like lakes and rivers. A survey has identified that almost over 92% of the toxic weighted pound or its equivalent can be released from hydrogen sulphide, dioxin and its related compounds as well as from manganese. (Taking Stock, 1999) In this regard, precautionary measures can be implemented like recycling and burning the waste, using bioremediation ponds as well as by employing less damaging agents.

## WASTE

Waste that are incinerated or commingled with groundwater via traditional burial methods, can be the additional hazard of toxic inks, dyes and polymers, which are identified to be the potential carcinogens and hence to be avoided.

# BLEACHING OF WOOD PULP: ENVIRONMENTAL CONSIDERATIONS

### *Chlorine and Chlorine Based Materials*

Chemical bleaching of wood pulp produced especially by the kraft and/ or sulfite process can be accomplished with the help of chlorine and related compounds. Natural elemental chlorine obtained from plants can significantly produce dioxin with sufficient quantities; the process can be significantly controlled with the replacement of ECF, i.e. Elemental Chlorine free and TCF, i.e. Totally chlorine free bleaching processes.

The process of TCF bleaching involves removal of chlorine and reduction of chlorinated organic compounds to the minimum levels;

while modern ECF can reduce AOX emission to less than 0.05 kg per tonne of pulp produced.

By the year, 2005 the use of elemental chlorine has been reduced to over 90% and the same is being replaced with kraft pulp made from ECF and TCF. A recent study in this regard, from US environmental protection agency concluded that ECF and TCF pulp produced due to oxygen delignification and extended delignification can be helpful in minimizing environmental effects. (Paper Task Force, 1995)

While there is no much of environmental difference observed between both ECF and TCF emission; when conventional reviews are compared with modern evidences, it has been evidently suggested that TCF waste are less toxic than their other counterparts. (Tarkpea et.al 1999)

## Kraft process and Sulfite process

The kraft process and the sulfite process involves application of uses Sulfur-based compounds for preparing wood pulp; however, it should be understood that Sulfur dioxide is water-soluble and a primary reason for acid rain fall.

Cause of the odor characteristic of pulp mills utilizing the kraft process are air emissions of hydrogen sulfide, methyl mercaptan, dimethyl sulfide, dimethyl disulfide, and other chemicals that are released into the air and water from most paper mills include the following (Bhatia et al,, 1978): carbon monoxide, ammonia, nitrogen oxide, mercury, nitrates, methanol, benzene, volatile organic compounds, and chloroform.

In bleaching mechanical pulp the organic material is retained, and the chemicals used (hydrogen peroxide and sodium dithionite) produce benign byproducts, those are not a key environmental concern. Though, the bleaching of chemical pulps releases organic materials into waterways cause significant environmental damage.

As evident by the report suggested by environmental organizations like Greenpeace, an increased public awareness towards environmental issues from the 1970s and -80s was observed, which in turn influenced the pulping industry and governments, especially on Swedish and Finnish pulp and paper companies to address the release of these materials into the environment (Tilman, 2008; Auer,1996).

Since, dioxins are known to accumulate in the food, they can be very hazardous to mankind as over 90% of human exposure is through food, primarily meat, dairy, fish and shellfish. (Radenkova-Saeva, 2009).

## GREENHOUSE GAS EMISSIONS

Globally, print and paper industry accounts for about 1% of carbon dioxide emissions, 69% of GHGs come from the energy and transportation industries. Since, paper is being disposed of, in the landfill sites it can be dangerously add on the environmental pollution, by subsequently breaking down in methane; thus, paper recycling can be healthy alternative in reducing global warming by almost 15-25%.

Accordingly, reports suggested that mainly due to self-generation of power from renewable biomass and reduced use of forest products; a steady decline in GHG emissions by 56% and total energy use by 30% in Canada, between 2000 and 2012, (Natural Resources Canada, 2016).

## CHEMICAL BLEACH: EFFECTS ON HEALTH

The potential health effects of bleach, have been listed by EPA, and can be noted as:

Poisonous gas created through mixing with ammonia, ammonium quaternary compounds and other acidic products. This can be corrosive to eyes, skin and respiratory irritant.

Contaminated air when inhaled for a longer period of time can impose a potential risk to many system, such as cardiovascular, gastrointestinal or liver, kidney, central nervous system, respiratory, and skin or sense organs.

Release of these chemicals can be toxic to aquatic organisms and should be used only with increased ventilation.

### Case Study

Cotton fiber is a single biological cell, built up of lumen, secondary wall, primary wall and cuticle. The inner protective cuticle is composed of pectin, fatty acids, wax, mineral matters, high molecular weight alcohols and their esters (Hossain and Uddin, 2011).

***Pretreatment***: Textile materials possess some natural or inheriting, or may be added purposely to be spun or woven better. These materials can sometimes be contaminated through accidental incorporation of impurities while their usage. Wherein, preparative processes, that are used mainly for the removal of these impurities before actual dyeing or printing can be advantageous and further classified into two main categories, as mentioned below:

*Cleaning processes*: the process through which impurities are removed by either physical or chemical parameters is known to be as the chemical process.

*Whitening processes*: The process that promotes optical whitening of the material for the destruction of traces of coloring matters is known to be as the whitening process.

***Conventional Process***: The first step of wet pre-treatment is the **desizing process,** for better weavability removal of starch or other sizing materials, which are applied on yarn before weaving. One of the widely used method of desizing is enzymatic desizing of cellulose fabrics, wherein natural as well as modified enzymes, like amylolytic enzymes are used to any form of starch into its water-soluble counterpart, without affecting the structural integrity of its fibers. Scouring is identified to be as the most important step, allowing complete as well as the partial removal of non-cellulosic components, along with the impurities incorporated through machinery and size lubricant in the native cotton. Traditionally, the process of scouring is achieved through a series of chemical treatments, the traces of which are subsequently removed through rinsing with water, caustic soda and detergent. This treatment apparently involves production of large amount of salts, acids and alkali that need large water to be removed.

Bleaching with $H_2O_2$ is effective on regenerated cellulose fibres and are widely applicable for the natural cellulosic fibres, protein fibres and a blend of synthetic as well as natural fibres, for e.g. cotton polyester. The maximum bleaching activity from $H_2O_2$ is obtained generally at about alkaline conditions pH 11.5 (Hossain and Uddin, 2011).

# THE GREENER APPROACH

Bioprocess technology today is based on the principle of combining whole organisms or enzymes with nutrients under the conditions to make the desired end product. Conventionally, in textile application malt was used to remove gum from fabrics before their printing; however, by the year 1857, the knowledge of specific enzymatic actions like amylase used for starch splitting, was being applied in practice.

Conventionally, in case of stone washing of many denim fabrics, it was evident that the color of indigo or Sulphur dyed material used to get faded; the issue can be solved with the incorporation of bio-stoning. It is found to be important in achieving a great washed-down or worn look to the new casuals. It is highly supported due to its easy environmental acceptance and decrease use of pumice stone that may be harmful to both machineries as well as fabric.

Thus, it has been evident that since bioprocesses are living material, which need optimum temperature, pressure and pH conditions; they can utilize renewable resources for the production of greater quantities and consumption of lesser energy.

***Enzymatic scouring and bleaching***: Compared to conventional alkaline boiling off, the advantages of bioscouring are to be considered as follows:

It utilizes only one rinsing cycle and thus can save water as well as time

It helps in saving energy consumption by lowering the treatment temperature to around $50^0$C (Rocky, 2012)

It allows less fibre weight loss along with lesser COD and BOD in the effluent.

Thus, in general the bleaching enzyme allows performance of the system at much lower temperatures than specified; thus, accounting for saving water and energy consumption of up to 40%

The different types of enzymes and their uses are mentioned in detail in Chapter 2: Enzymes / Microbes used for bleaching.

# MECHANISM OF HYDROGEN PEROXIDE BLEACHING

Hydrogen peroxide can be referred to be a very weak acid; which could be ionized to form perhydroxyl ions ($HOO^-$) $H_2O_2$ $HOO^-$ + $H^+$; the formation of which, is in turn favored by the alkaline conditions. The bleaching is caused by the anions which are the source of active oxygen. $H_2O_2$ + $OH^-$ $H_2O$ + $HOOHOO^-$ $OH^-$ + [O]

Although, the exact mechanism of bleaching cannot be understood, being very complicated; different theories have been put forth to justify it. One hypothesis is that the color producing agents in natural fibers are organic compounds containing conjugated double bonds.

## *Case Study 1*

*Invatex LTA:* It can be referred to be as the enzyme that is primarily used to assist and boost peroxide reaction in the gentle powder bleach, which may catalyze the peroxide bleaching in combination with Invatx LTA. (Hossain and Uddin, 2011). Grey woven fabric – cotton (plain weave) and grey knit fabric – viscose were the materials used.

*Bio 6*: on cotton shows good absorbency, but is somewhat uneven. In case of viscose, absorbency of fabric treated conventionally and with bio-process is same.

*Strength loss*: Bio process is found to be highly efficient in strengthening the fibers as compared to the conventional process.

*Dye uptake %:* It was found that more dyes are absorbed in cotton treated by bioprocess than those treated by conventional process, which did not correspond to the shade %.

*Whiteness index*: It can be concluded from accumulated data that the whiteness index is highest for the fabric processed with 9 g/l $H_2O_2$ in the conventional way. Contrary to which, for viscose, the whiteness index achieved with all the processes are almost the same. As a matter of fact, bioprocesses can give excellent results as compared to the conventional methods.

## *Case Study 2*

Wheat straw is non-wood agricultural waste portion as an excellent fiberous material to recycle into quality paper (Yaqoob et al,, 2011). Celluloses and lignin are two major constituents, the chemical composition of which vary according to the wheat straw genetics, geographic conditions1 and/ or pulping/cooking condition to pulp straw. Other constituents are the silica, ashes and extractive.

The main purpose of bleaching pulp is to increase its brightness level to obtain white paper. The bleaching process is either to stabilize or solubilize to wash away the lignin contents while cellulose contents remains intact in the pulp fiber. Thus, bleaching can be done by (i) Delignification: Removing the residual lignin of chemical pulps. (ii) Lignin-conserving or holding bleaching: without loss of substance chromophoric groups of mechanical pulp are converted or stabilized. Delignifying bleaching chemicals can be classified into three groups.

*Group I*: chemicals of group I react with all aromatic lignin units, with phenolic groups and their double bond (Fig. 1) e.g., chlorine, ozone, peroxy acids, etc.

*Group II*: chemicals of group II reacts mainly with free phenolic hydroxyl group (Fig. 1) e.g., oxygen, chlorine dioxide.

*Group III*: chemicals of group III reacts mainly with certain functional conditions groups in lignin, e.g., carbonyl group (Fig. 1) e.g., hydrogen peroxide in alkaline media. These chemicals proved to be highly effective to delignify and/or brighten various pulp varieties. But these chlorine-containing chemicals once released into environment could be strong cancer-causing.

Chen et al,(1999) reported another total chlorine free agent potassium permanganate as intermediate bleaching stage between oxygen and peroxide. They worked out OMnP bleaching sequence for soda-AQ pulp. Their work also illustrates that chelation prior to peroxide treatment is helpful to increase brightness to 80 % ISO.

Hong et al, (, adopts the method of enzymatic pre-treatment using xylanase prior to bleaching for improving bleachability and brightness of chemo-mechanical wheat straw pulp. The increased pulp brightness even at 50 % reduced dosage of hydrogen peroxide proved that xylanase pre-treatment assisted to improve the bleachability of straw pulp.

Han et al, (2001), worked out enzymatic applications on increased bleachability of straw pulp. They devised a bio-bleaching laccase-mediator enzyme system followed by peroxide treatment. Use of laccase-mediator system not only reduces the consumption of bleaching agents but can also help to reduce the load of wastes on environment with a negligible loss of pulp strength.

The application of enzyme and ozone in a multistage XOAZRP sequence to fully bleach the wheat straw pulp was defined (Roncero et al,, 2003).

Auto-catalyzed ethanol wheat straw pulp was examined and they remained successful in developing two bleaching sequences without any loss of strength. Cao et al, (2005) cooked wheat straw kraft pulp at low lignin contents for total chlorine free bleaching ans found that AQ(PN)P and QZEYP bleaching sequences are successful in achieving brightness greater than 80 % SBD, suggested that cellulose viscosity of bleached pulp and its brightness stability is better than chlorinated bleached pulp using CEH sequence.

Although the chemical composition of wheat straw pulp and its pulping conditions may vary according to location and its genetic makeup, literature reveals chlorine free bleaching sequences for pulp fibers originated and cooked at varied conditions which offer much smaller environmental footprints than conventional bleaching methodologies.

## LIMITATIONS

Since scouring chiefly targets pectinase containing non-cellulosic impurities, the process is technically feasible; however, the reason for its sparse use in the dye industry is still not understood. Accordingly, some of the possible limitations are discussed herewith:

It has little or no effect on mote removal; although the process mainly is applicable for darker shades, it cannot prevent the appearance of motes in darker fabrics.

The inability of bioprocess to achieve better whitening can be considered as another important limiting factor. The process hence cannot be used to treat pale, medium shade and fully white colored fabrics.

Lastly, inability to remove wax impurity can be attributed towards its limited acceptance.

## THE FUTURE

An important source of problems for human and environmental health is release of improperly treated textile wastes into the environment. The textile industry is also focused due to the large volumes of water used by its industrial park, consequently generating large volumes of waste for each kilogram of cotton produced approximately 200 liters of water are needed. These wastes, ranging from original colors that are lost during the dyeing process to different pesticides associated with the process are complex mixtures of many pollutants that can add on serious contamination of the fresh water resources (Farah MDC, et al, 2013). Moreover, the biological cycles of the aquatic biota changes due to the pollution of water bodies with these compounds; the effect on the water bodies is more pronounced in photosynthesis and oxygenation processes.

Moreover, studies have shown that some classes of dye, especially azo dyes and their by-products, may be carcinogenic and / or mutagenic (Carneiro et al, 2010; Ferraz et al,, 2011), endangering human health, since the wastewater treatment systems and water treatment plants (WTP) are ineffective in removing the color and the mutagenic properties of some dyes (Oliveira et al,, 2010).

Thus, these highly stable compounds are very difficult to be removed easily from environments and are further designed to resist biodegradation to meet the increasing demands of durability.

Many federal countries, such as the United States of America, Canada and Australia have national environmental legislation, which, as in Europe, establishes the limits that must be complied with. In some countries, for example India, Pakistan and Malaysia, the emission limits are recommended, but are not mandatory (Hessel et al,, 2007).

## CONCLUSION

Considering the utmost necessity for upgradation of bioprocesses over conventional methods, mainly due to elevated fuel pricing, electricity shortage and increasing demand of natural resources; it is important to move ahead with sustainable development. Although, it has been apparent that the enzymatic processes or bio processes for that matter can definitely be considered as a break-through towards a greener technology; more research is still required to resolve some of the underlying technical issues.

## REFERENCES

1. Auer MR (1996) Negotiating toxic risks: A case from the Nordic countries. Environ Politics 5:687-699.
2. Bahrum AMK, Prang R (2012) Comparison of Effectiveness Between Conventional Scouring & Bio-Scouring On Cotton Fabrics. Int J Sci Eng Res 3(8).
3. Bajpai P (2004) Biological bleaching of chemical pulps. Crit Rev Biotechnol 24(1):1-58.
4. Bhatia SP, de Souza TL, Azarniouch MK, Prahacs S (1978) Air pollution control in kraft pulp mills. Am Ind Hyg Assoc J 39(2):83-93.
5. Carneiro PA, Umbuzeiro GA, Oliveira DP, Zanoni MVB (2010) Assessment of water contamination caused by a mutagenic textile effluent/dyehouse effluent bearing disperse dyes. J Hazard Mater 174(1-3):694-699.
6. Environment and Climate Change Canada. 2016. http://www. ec.gc. ca/pollution/default.asp?lang=En&n=E96450C4-1
7. Environmental Impact of Paper. http://www.econatics.co.za/?page_ id=19899Environmental impact of printing inks. (March 2013). http://www.eupia.org/uploads/tx_edm/2013-03-05_EuPIA_ Environmental_Impact_of_Printing_Inks.pdf
8. EPA. Fact Sheets–National Emission Standards for Hazardous Air Pollutants (NESHAP) from the Pulp and Paper Industry (40 CFR 63, Subpart S). Stationary Sources of Air Pollution. 2016.
9. EPA. National Emission Standards for Hazardous Air Pollutants for Source Category: Pulp and Paper Production; Effluent Limitations Guidelines, Pretreatment Standards, and New Source Performance Standards: Pulp, Paper, and Paperboard Category. Federal Register, 63 FR 18504, 1998-04-15. Biobleaching 88
10. EPA. Pulp, Paper, and Paperboard and the Builders› Paper and Board

Mills Point Source Categories Effluent Limitations Guidelines, Pretreatment Standards, and New Source Performance Standards. Federal Register, 47 FR 52006, 1982-11-18.

11. Farah MDC, et al (2013) Textile Dyes: Dyeing Process and Environmental Impact, Eco-Friendly Textile Dyeing and Finishing, Dr. Melih Gunay (Ed.), InTech, DOI: 10.5772/53659. Available from: https://www.intechopen.com/books/eco-friendly-textile-dyeingand-finishing/textile-dyes-dyeing-process-and-environmentalimpact
12. Ferraz ERA, Umbuzeiro GA, de-Almeida G, Caloto-Oliveira A, Chequer FMD, Zanoni MVB, et al (2011) Differential Toxicity of Disperse Red 1 and Disperse Red 13 in the Ames Test, HepG2 Cytotoxicity Assay, and Daphnia Acute Toxicity Test. Environ Toxicol 26(5):489-497
13. Gray WB, Shadbegian RJ (2015) Multimedia Pollution Regulation and Environmental Performance: EPA's Cluster Rule. RFF DP 15-26
14. Hessel C, Allegre C, Maisseu M, Charbit F, Moulin P (2007) Guidelines and legislation for dye house effluents. J Environ Manage 83(2):171-180. DOI: 10.1016/j. jenvman.2006.02.012.
15. Hong F, Liu S, Gui R, Shen ZB. Chem. Ind., 21 (2001)
16. Hossain S, Uddin MK (2011) Comparative Analysis Between Conventional Pretreatment and Bio-Preparation. Int J Eng Technol 11(3).
17. Chen J, Yang G, Li Z, Xuebao ZZ (1999) 14,51.
18. Radenkova-Saeva J (2009) Immunotoxicity of Dioxins. J Clin Med 2(2):8-13
19. Kristina S, Ann-Kristin B (2013) Growing Green and Competitive—A Case Study of a Swedish Pulp Mill. Sustainability 5(5):1789-1805; doi:10.3390/su5051789
20. Cao M, Tong G, Xue G, Xianweisu Kexue Yu Jishu XKY. (2005) Biobleaching: Status, Roadblocks and Future. 13(7).
21. Roncero MB, Torres AL, Colom JF, Vidal T (2003) Bioresour Technol 87:305.

22. Natural Resources Canada. 2016. The State of Canada's Forests. Annual Report 2015

23. Oliveira GAR, et al (2010) Chlorination treatment of aqueous samples reduces, but does not eliminate, the mutagenic effect of the azo dyes Disperse Red 1, Disperse Red 13 and Disperse Orange 1. Mutat Res 703(2):200-208.

24. Palmer K, Oates WW, Portney P (1995) Tightening Environmental Standards: The Benefit Cost or the No-Cost Paradigm? J Econ Perspect 9:119–132.

25. Paper Task Force (1995). Environmental Comparison of Bleached Kraft Pulp Manufacturing Technologies. White paper no. 5. Joint publication of Duke University, Environmental Defense Fund, Johnson & Johnson, McDonald›s, Prudential Insurance Company of America and Time Inc.

26. Taking stock 1999. North American pollutant releases and transfers. Commission for Environmental Cooperation of North America 2002. http://www3.cec.org/islandora/en/item/1811-taking-stock-1999-north-american-pollutant-releases-and-transfers-en.pdf

27. Tarkpea M, et al (1999) Toxicity of conventional, elemental chlorine–free, and totally chlorine–free kraft-pulp bleaching effluents assessed by shortterm lethal and sublethal bioassays. Environ Toxicol Chem 18(11):2487–2496.

28. Tilman A (2008) Pulp and Paper Pollution: The Toxic Legacy of Federal Neglect. Archived May 24, 2013, at the Wayback Machine. Reach for Unbleached Foundation, Comox, BC.

29. Yaqoob N, Kausar JC, Bushra M (2011) Review on Total Chlorine Free Bleaching Sequences of Wheat Straw Pulp. Asian J Chem 23(8):3317-3319.

CHAPTER 6

# DEGRADATION OF TEXTILE DYES USING IMMOBILIZED LIGNIN PEROXIDASE-LIKE METALLOPORPHINES UNDER MILD EXPERIMENTAL CONDITIONS

**Paolo Zucca[1,2], Antonio Rescigno[1], Manuela Pintus[1] , Andrea C Rinaldi[1] and Enrico Sanjust[1]**

[1] Dipartimento di Scienze Biomediche, Università di Cagliari, Cittadella Universitaria, 09042, Monserrato, CA, Italy

[2] Consorzio UNO (University of Oristano Consortium), Oristano, Italy

## ABSTRACT

### Background

Synthetic dyes represent a broad and heterogeneous class of durable pollutants, that are released in large amounts by the textile industry. The ability of two immobilized metalloporphines (structurally emulating the ligninolytic peroxidases) to bleach six chosen dyes (alizarin red

S, phenosafranine, xylenol orange, methylene blue, methyl green, and methyl orange) was compared to enzymatic catalysts. To achieve a green and sustainable process, very mild conditions were chosen.

## Results

IPS/MnTSPP was the most promising biomimetic catalyst as it was able to effectively and quickly bleach all tested dyes. Biomimetic catalysis was fully characterized: maximum activity was centered at neutral pH, in the absence of any organic solvent, using hydrogen peroxide as the oxidant. The immobilized metalloporphine kept a large part of its activity during multi-cycle use; however, well-known redox mediators were not able to increase its catalytic activity. IPS/MnTSPP was also more promising for use in industrial applications than its enzymatic counterparts (lignin peroxidase, laccase, manganese peroxidase, and horseradish peroxidase).

## Conclusions

On the whole, the conditions were very mild (standard pressure, room temperature and neutral pH, using no organic solvents, and the most environmental-friendly oxidant) and a significant bleaching and partial mineralization of the dyes was achieved in approximately 1 h. Therefore, the process was consistent with large-scale applications. The biomimetic catalyst also had more promising features than the enzymatic catalysts.

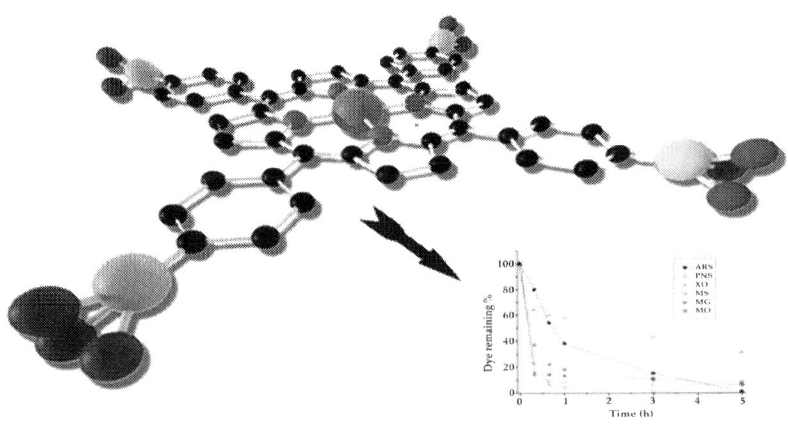


**Keywords:** Biomimetic, Dye, Lignin-peroxidase, Porphine, Porphyrin

# BACKGROUND

Textile dyes are a broad and heterogeneous class of molecules used in many technological fields (i.e., textile and paper production, food technology, and hair coloring) [1]. Many chemical classes of synthetic dyes are frequently employed on an industrial scale; azo dyes, triarylmethanes, phenothiazine, and anthraquinones are among the most widespread [2, 3].

Because the goal of the dyeing process is color durability, textile dyes of all chemical classes are commonly highly resistant to both chemical and physical degradation. Accordingly, textile dyes represent a serious environmental concern because of their chemical inertness, high annual production (over 10,000 tons per year [1]), toxicity/carcinogenicity [4, 5], and the large amount of dyes released in industrial wastewaters (up to 50% of total process intake [6]).

The European Directive 2002/61/EC forbids the use of some products (in particular some azo dyes). However, these restrictions only marginally solve the environmental problem because of the large volume of other dyes discharged every year.

Several physical, chemical and enzymatic approaches have been proposed to remove dyes from industrial wastewaters [7, 8, 9, 10, 11, 12, 13, 14, 15, 16] that require long reaction times, extreme operational conditions, and high costs [1]. Accordingly, the development of a clean and environmentally friendly process for removing dyes is a challenging issue, far from being solved.

Redox-active metalloporphines are efficient catalysts; in the presence of a proper oxidant [17, 18, 19, 20, 21], they are able to oxidize many substrates, including several textile dyes, under very mild conditions. We previously studied the ability of an immobilized manganese porphine to oxidize alizarin red S (ARS), an anthraquinone dye) [22], and phenosafranine (PNS), a phenylphenazinium dye [23], elucidating several aspects of the catalytic mechanism.

In this study, we compare the catalysts IPS/MnTSPP and PP-PVA/FeTFPP (Figure 1), each containing a metalloporphine (Mn- and Fe-, respectively) immobilized onto a solid support, thus emulating ligninolytic peroxidases [24]. We investigated the ability of these two catalysts to oxidize different chemical classes of textile dyes in the presence of hydrogen peroxide under mild conditions. We also report the results of these catalysts on ARS and PNS to allow a comprehensive comparison. All the dyes included in the study are reported in Table 1.

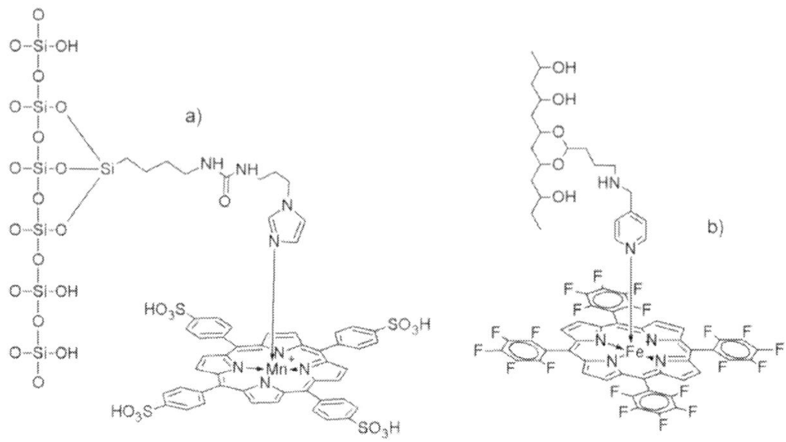


**Figure 1:** Proposed structure of the catalysts IPS/MnTSPP (a) and PP-PVA/FeTFPP (b). For more details, see: [24, 25].

**Table 1:** The six textile dyes included in this study

| **Dye** | | **MW (g/mol)** | $\lambda_{max}$**(nm)** | **ε (mol * L$^{-1}$ * cm$^{-1}$)** |
|---|---|---|---|---|
| Alizarin Red S (ARS) | 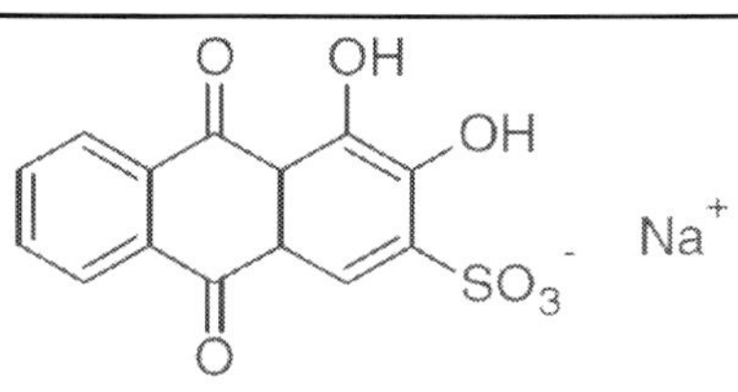 | 342.26 | 520 | 7200 |
| Pheno-safranine (PNS) | 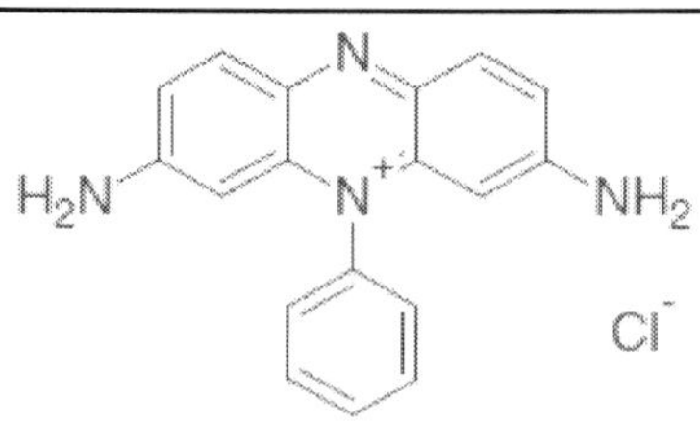 | 322.80 | 520 | 32000 |

| Xylenol Orange (XO) | 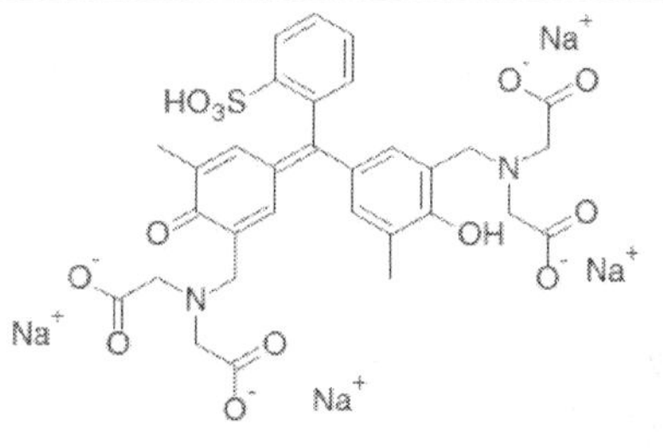 | 782.56 | 435 | 10500 |
|---|---|---|---|---|
| Methylene Blue (MB) | 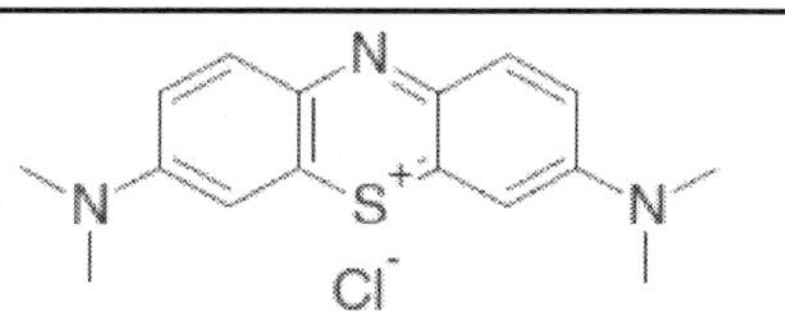 | 319.86 | 660 | 79500 |
| Methyl Green (MG) | 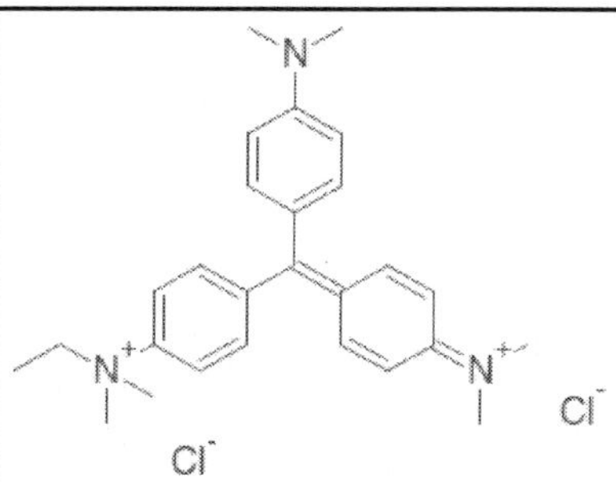 | 472.51 | 632 | 55700 |
| Methyl Orange (MO) | 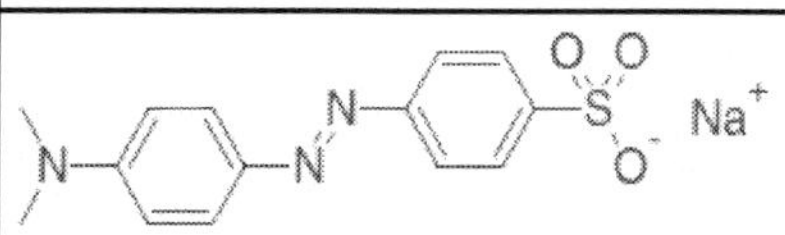 | 327.34 | 464 | 24800 |

# RESULTS AND DISCUSSION

## Comparison of IPS/MnTSPP and PP-PVA/FeTFPP

We previously developed two biomimetic catalysts, IPS/MnTSPP and PP-PVA/FeTFPP, which are able to oxidize several lignin-model compounds. IPS/MnTSPP strictly emulates a lignin peroxidase (LiP) active site, having an imidazole-grafted support that coordinates a Mn(III)-porphine [24]. This catalyst showed a LiP-like activity, whereas, unlike an enzyme such as manganese peroxidase (MnP), it was not able to use the $Mn^{2+}$ ion as a redox mediator. Conversely, PP-PVA/FeTFPP was only “bioinspired” (i.e., only inspired by but not strictly emulating, a natural enzyme active site) because the metalloporphine was coordinated by a pyridine-grafted support [25] rather than imidazole, as in ligninolytic peroxidases. This catalyst showed both LiP-like and MnP-like catalytic activities.

These two catalysts were tested for their ability to decolorize several textile dyes in the presence and absence of $Mn^{2+}$ using hydrogen peroxide as the oxidant in accordance with green chemistry principles [26]. The investigation employed very mild conditions (neutral pH and room temperature, using no organic solvents, and the most environmental-friendly oxidant) to serve as the basis for an environmentally sustainable textile wastewater treatment process.

The results of these tests are reported in Table 2. Both biomimetic catalysts easily degraded all tested dyes, showing wide substrate specificity. As previously reported [24], the catalytic behavior of IPS/MnTSPP was not affected by the presence of Mn(II), showing similar bleaching percentages in both the presence and absence of this ion. All six dyes were rapidly bleached under the conditions studied; more than with a removal in all cases higher than 50% of each dye (up to > 90% for PNS, MG, and MO) was removed in 1 h. In contrast, PP-PVA/FeTFPP was almost inactive in the absence of a suitable redox mediator. The conversion of all six dyes increased in the presence of $Mn^{2+}$, leading to values similar on average to those observed for IPS/MnTSPP.

**Table 2:** Removal of the studied dyes in the presence of hydrogen peroxide and either IPS/MnTSPP or PP-PVA/FeTFPP, in both the presence and absence of Mn(II) as the redox mediator

| **Mn(II)** | **Catalyst** | **Dyes removal % (1 h)** | | | | | |
|---|---|---|---|---|---|---|---|
| | | **ARS** | **PNS** | **XO** | **MB** | **MG** | **MO** |
| No | IPS/MnTSPP | 62 ± 1 | 92 ± 2 | 42 ± 4 | 96 ± 4 | 82 ± 5 | 87 ± 2 |
| | PP-PVA/FeTFPP | n.d. | n.d. | n.d. | n.d. | n.d. | n.d. |
| Yes | IPS/MnTSPP | 55 ± 1 | 88 ± 2 | 44 ± 2 | 91 ± 3 | 77 ± 4 | 85 ± 4 |
| | PP-PVA/FeTFPP | 48 ± 3 | 58 ± 2 | 55 ± 5 | 37 ± 1 | 48 ± 2 | 91 ± 2 |

The reaction mixtures contained 5 mg of catalyst in 1 mL of 25 mM buffer solution containing 8.8 mM hydrogen peroxide and a proper concentration of the selected dyes (0.29 mM ARS, 0.31 mM PNS, 2 mM XO, 0.15 mM MB, 1.5 mM MG, and 1.25 mM MO). Quantification of the removed dye was obtained photometrically, using the $\lambda_{max}$ and molar extinction coefficients reported in Table 1. n.d.: not detectable. ($n = 4$).

Considering that textile wastewaters do not usually contain manganous ion and that it would therefore be necessary to add $Mn^{2+}$, itself a heavy metal pollutant, to the reaction, IPS/MnTSPP was considered the most promising catalyst for future industrial applications Hence, the rest of study focused on IPS/MnTSPP.

The time dependence of dye bleaching by IPS/MnTSPP is shown in Figure 2. Note that most of the bleaching occurs in the first hour of the reaction, consistent with industrial timing needs.

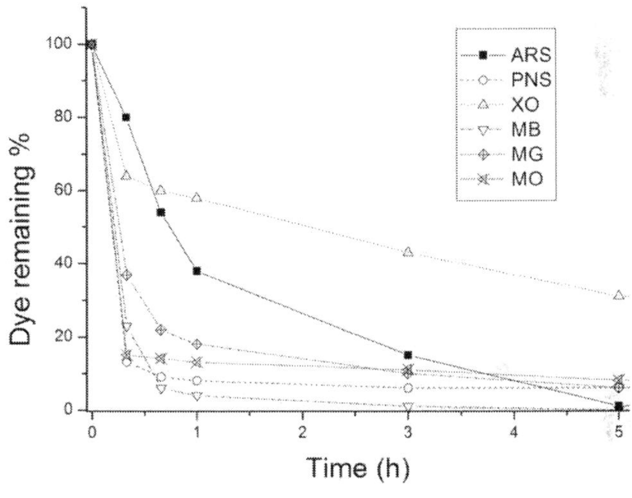


**Figure 2:** Time dependence of dye removal in the presence of IPS/MnTSPP and hydrogen peroxide. The reaction mixtures contained 5 mg of catalyst in 1 mL of 25 mM buffer solution containing 8.8 mM hydrogen peroxide and a proper concentration of the selected dyes (0.29 mM ARS, 0.31 mM PNS, 2 mM XO, 0.15 mM MB, 1.5 mM MG, and 1.25 mM MO). Quantification of the dyes was obtained photometrically, using the $\lambda_{max}$ and molar extinction coefficients reported in Table 1.

## Characterization of Bleaching by IPS/MnTSPP and $H_2O_2$

IPS/MnTSPP catalysis was strongly pH-dependent; an approximately neutral pH was optimal for all dyes tested except XO and PNS, which had optimal bleaching at pH 6 and 8, respectively (data not shown). However, XO and PNS were sufficiently bleached at pH 7 that IPS/MnTSPP remains a viable catalyst for environmentally friendly wastewater treatment.

An exhaustive kinetic modeling of this catalysis is beyond the scope of our work. However, the decolorization of all six dyes apparently followed saturation Michaelis-Menten-like kinetics, allowing the calculation of catalytic parameters (Table 3). Reported Michaelis constants ($K_M$) showed that IPS/MnTSPP had significantly higher affinity for cationic dyes than anionic dyes (lower $K_M$ for PNS and MB, higher $K_M$ for ARS and XO). This lower affinity led to the lowest bleaching rate for XO and ARS (compare Figure 2). Notably, in all cases, $K_M$ values were in the millimolar range and did not suffer from substrate inhibition. This can be a crucial issue for potential industrial applications because some bleaching approaches (i.e., enzymatic, *vide ultra*) are not able to work at these concentrations.

**Table 3:** Kinetic parameters measured for the bleaching of each dye by IPS/ MnTSPP and hydrogen peroxide

| Sub-strate | Kinetic parameter | Dye | | | | | |
|---|---|---|---|---|---|---|---|
| | | ARS | PNS | XO | MB | MG | MO |
| Reducing substrate | $K_M$ (mM) | 2.11 ± 0.74 | 0.32 ± 0.04 | 1.33 ± 0.38 | 0.11 ± 0.02 | 0.72 ± 0.12 | 0.63 ± 0.05 |
| | $k_{cat}$ (min$^{-1}$) | 0.14 ± 0.03 | 2.2 ± 0.1 | 98 ± 14 | 19.0 ± 1.8 | 108 ± 9 | 67 ± 4 |
| | $k_{cat}/K_M$ (mM$^{-1}$ min$^{-1}$) | 0.066 ± 0.037 | 7.0 ± 1.2 | 74 ± 31 | 166 ± 52 | 150 ± 37 | 105 ± 16 |
| $H_2O_2$ | $K_M$ (mM) | 3.21 ± 0.12 | 3.0 ± 0.4 | 7.9 ± 1.9 | 2.4 ± 0.5 | 0.43 ± 0.08 | 7.4 ± 1.5 |
| | $k_{cat}$ (min$^{-1}$) | 0.13 ± 0.01 | 1.9 ± 0.1 | 78 ± 8.9 | 15.2 ± 1.3 | 159 ± 12 | 71 ± 8 |
| | $k_{cat}/K_M$ (mM$^{-1}$ min$^{-1}$) | 0.040 ± 0.004 | 0.63 ± 0.13 | 9.9 ± 3.6 | 6.4 ± 2.1 | 369 ± 100 | 9.6 ± 3.2 |

Data for ARS and PNS were previously reported and are added here to allow a full comparison [22, 23] ($n = 5$).

To test the hypothesis of a one-electron abstraction mechanism as the first step of the bleaching process, the influence of several well-known redox mediators was also studied: 4-hydroxy-2,2,6,6-tetramethylpiperidine-1-oxyl (OH- TEMPO), *N*-hydroxyphthalimide

(NPH), *N*-hydroxysuccinimide (NHS), and *N*-hydroxybenzotriazole (NHT). The results are summarized in Table 4. Only in the case of MG was an increase (three-fold) of the catalytic activity observed; in all the other cases, the redox-mediators mainly decreased IPS/MnTSPP catalysis (this quenching could be due to the lower reactivity of the oxygen-centered radical form of these mediators). Even in the best case, however, because the increase in catalytic activity was moderate, the industrial benefit of these redox mediators may not justify their environmental impact.

**Table 4:** Effect of well-known redox-mediators on the bleaching of the six tested dyes by IPS/MnTSPP

| **Redox mediator** | **% Catalytic activity** | | | | | |
|---|---|---|---|---|---|---|
| | **ARS** | **PNS** | **XO** | **MB** | **MG** | **MO** |
| OH-TEMPO | 98 | 45 | 277 | 56 | 336 | 16 |
| NHT | 142 | 46 | 162 | 38 | 299 | 121 |
| NHS | 141 | 89 | 146 | 90 | 313 | 122 |
| NPH | 103 | 32 | 74 | 56 | 340 | 57 |

Results are expressed by referring to a 100% reference activity in the absence of any mediator. Data for ARS and PNS were previously reported and are added here to allow a full comparison [22, 23].

The catalyst was also investigated for its reuse potential. The results are reported in Table 5. On average, a rapid decrease of catalytic activity was observed in the first 2–3 cycles. Subsequently, activity was stable at approximately 25-30% of maximum for at least 8–10 cycles. A potential explanation for this pattern is the possibility that a fraction of some dyes remains tightly adsorbed on the catalyst even when the supernatant is completely decolorized. This adsorbed dye would likely reduce the catalytic performance by steric hindrance and/or electrostatic repulsion of incoming substrate molecules. Nevertheless, the catalyst keeps a significant part of its catalytic activity during multi-cycle employment. This is an important quality for large-scale applications.

**Table 5:** Multi-cycle activity of IPS/MnTSPP

| Catalytic cycle | % Catalytic activity | | | | | |
|---|---|---|---|---|---|---|
| | **ARS** | **PNS** | **XO** | **MB** | **MG** | **MO** |
| 1 | 100 | 100 | 100 | 100 | 100 | 100 |
| 2 | 68 | 73 | 100 | 99 | 98 | 44 |
| 3 | 49 | 70 | 93 | 73 | 97 | 37 |
| 4 | 33 | 61 | 88 | 49 | 97 | 34 |
| 5 | 30 | 51 | 47 | 37 | 97 | 28 |
| 6 | 28 | 49 | 41 | 34 | 96 | 28 |
| 7 | 28 | 44 | 31 | 34 | 96 | 28 |
| 8 | 26 | 41 | 30 | 26 | 94 | 27 |
| 9 | 25 | 39 | 29 | 25 | 94 | 20 |
| 10 | 25 | 35 | 27 | 24 | 90 | 19 |

Catalytic activity of the IPS/MnTSPP adduct upon multicyclic use during the decolorization of the studied dyes ($n = 3$). Data for ARS and PNS were previously reported and are added here to allow a full comparison [22, 23].

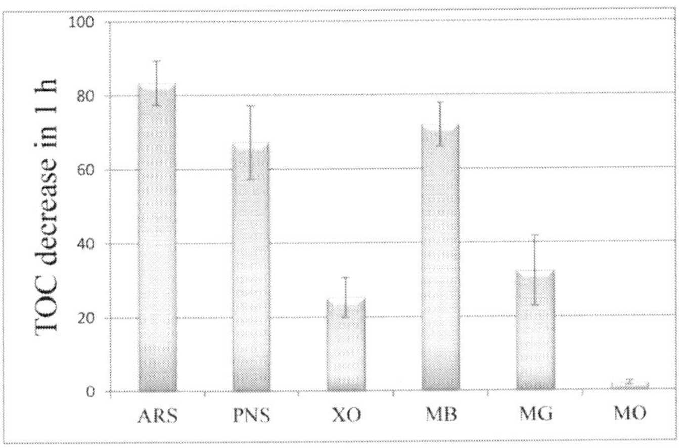


**Figure 3:** Total organic carbon (TOC) decrease during the bleaching of the textile dyes by IPS/MnTSPP ( $n = 3$).

Significant removal of total organic carbon (TOC) was also observed during the bleaching by $H_2O_2$/IPS/MnTSPP, as shown in Figure 3. In the case of ARS, PNS, and MB, over 50% of TOC removal was observed in 24 hours, whereas lower decreases were recorded for XO and MG. Conversely, in the case of MO no TOC modification occurred during the catalysis. These results suggest that a decarboxylative degradation occurred for some of the dyes, whose TOC values decreased significantly during the bleaching experiments.

IPS/MnTSPP was also tested for its ability to decolorize a complex mixture of all the six dyes, emulating realistic textile plant waste. The six dyes were each present in the saturating concentrations used above, leading to high impacting TOC (approximately 1580 mg/L). The changes in the UV–vis spectra (Figure 4) show a significant decolorization, although with a slower kinetics (approximately 5 hours for a 50% decrease of the main visible peak). However, the feasibility of using this treatment in an industrial process was confirmed by a significant TOC decrease (approximately 20%) in the same time frame.

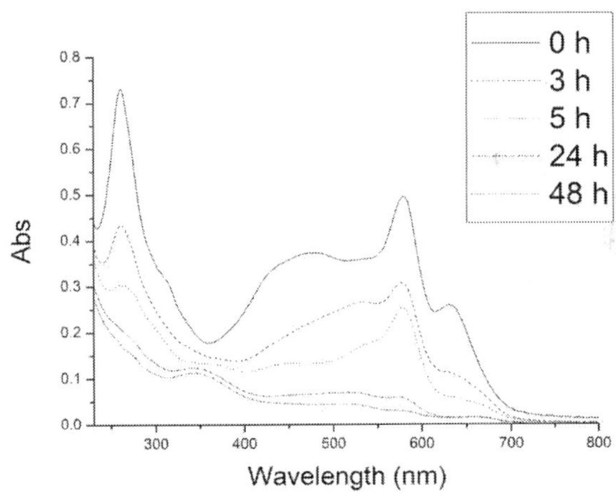


**Figure 4:** Changes in UV–vis spectra during the bleaching of a mix of the six studied dyes by IPS/MnTSPP.

## Comparison with Enzymatic Bleaching

A full comparison with the enzymatic bleaching of the dyes was also performed. The ligninolytic enzymes tested were laccase (LC), lignin peroxidase, and the non-ligninolytic enzyme horseradish peroxidase (HRP). In addition, Mn(III) was also used as oxidizing agent, mimicking manganese peroxidase action [27]. These biological catalysts suffered from significant substrate inhibition, preventing their use at the dye concentrations used for IPS/MnTSPP. Usually, concentrations of dyes at least of one order of magnitude lower were necessary. These enzymes also suffer from $H_2O_2$ inhibition, needing lower concentrations of oxidizing substrate. Accordingly, under the same conditions, it is not surprising that enzymatic bleaching efficiency was significantly lower than IPS/MnTSPP bleaching efficiency (<1% dye removal in all cases). Enzymatic bleaching activity, under the conditions studied, is reported in Table 6, allowing a qualitative analysis of substrate specificity only.

**Table 6:** Enzymatic bleaching of the dyes

| Dye | LC | LC + $H_2O_2$ | HRP | LiP | $Mn^{3+}$ |
|---|---|---|---|---|---|
| **ARS** | + | ++ | +/– | +/– | + |
| **PNS** | - | - | - | +/– | - |
| **XO** | + | + | + | - | + |
| **MB** | - | +/– | +/– | +/– | - |
| **MG** | + | + | ++ | +/– | ++ |
| **MO** | + | +/– | + | +/– | + |

In the conditions tested, ligninolytic enzymes present a narrower substrate specificity towards the dyes included in this study than IPS/MnTSPP. Legend: LC laccase, HRP horseradish peroxidase, LiP lignin peroxidase, ++ good substrate (complete bleaching < 1 h), + substrate, +/– weak substrate (bleaching < 10% in 24 h), - no substrate.

None of the examined enzymes were able to bleach all the studied dyes. For instance, PNS is resistant to LC, HRP, and $Mn^{3}$+, whereas only a partial degradation with concomitant irreversible enzyme inactivation is observed with LiP. MB underwent only partial oxidation with the enzymatic catalysts.

IPS/MnTSPP is able to effectively bleach all the dyes studied, as described above. This feature, combined with its ability to oxidize relatively high concentration of dye (in the millimolar range), and with its partial multi-cycling, may make IPS/MnTSPP catalysis a feasible alternative for industrial treatment of textile wastewaters.

# EXPERIMENTAL SECTION

## Chemicals and Instrumentation

All reagents were the best commercial grade available and were used without further purification. In particular, silica gel 100 came from Fluka (cat. No. 60746), and MnTSPP and FeTFPP were purchased from Sigma-Aldrich (cat. No. 441813 and 252913, respectively), PVA from Aldrich (cat. No. 363138), glutaraldehyde from Fluka (cat. no. 49629), horseradish peroxidase (E.C. 1.11.1.7) and lignin peroxidase (E.C. 1.11.1.14) were from Sigma-Aldrich (cat No. P-6782 and 42603, respectively). Laccase from *Pleurotus sajor-caju* was purified through two chromatographic steps (details in [28, 29]).

Spectrophotometric measurements were carried out with an UltroSpec 2100pro (Amersham Bioscience). Concentrations of the substrates were also measured by UV-HPLC and GC-MS, using previously described methods [30, 31].

## Preparation of IPS/MnTSPP

The complex IPS/MnTSPP was synthesized as previously described [24]. Briefly, 10 g plain silica gel was functionalized by overnight reaction at 80°C with 10 mmol 3-(1-imidazolyl)propylcarbamoyl-3'-aminopropyl-triethoxysilane solubilized in 10 mL dioxane, in the presence of 10 µL *N*-methylmorpholine acting as the catalyst. The activated silica (IPS) was washed consecutively with 0.5 M HCl, $H_2O$, 0.1 M NaOH and $H_2O$ again. The wet silica was then carefully dried overnight in a vacuum oven.

The bound MnTSPP was quantified by the difference in absorption at 400 nm (the $\lambda_{max}$ of the MnTSPP Soret band), as described in [24].

## Preparation of PP-PVA/FeTFPP

The complex PP-PVA/FeTFPP was synthesized as previously described [25]. Briefly, 1 g of aminopropyl crosslinked PVA (prepared as in [32]) was suspended in an excess of 0.05 M sodium acetate buffer pH 5, and treated with 0.1 mL 4-pyridinecarboxaldehyde and 0.5 g sodium cyanoborohydride. After 24 h of incubation at 25°C, the support was exhaustively washed with water, and dried at 50°C, to obtain PP-PVA. One-gram aliquots of PP-PVA were treated with 20 mg of FeTFPP, solubilized in 10 mL DMSO. The slurries were kept stirring in the dark for 24 h, washed exhaustively with DMSO, and carefully dried at 50°C.

Bound metalloporphine was quantified by difference through spectrophotometric measurement at 411 nm ($\varepsilon_{411}$ = 115,000 $M^{-1}$ $cm^{-1}$ in DMSO) [25].

## Activity Measurements

Routine measurements of the catalytic activity of immobilized metalloporphines were performed through spectrophotometric assays. The suspension contained 5 mg of catalyst in 1 mL of 25 mM buffer solution containing 8.8 mM hydrogen peroxide and a proper concentration of the selected dye. The routine analyses were performed using the following concentrations of dyes (chosen to be only slightly higher than respective $K_M$, to better observe influences on catalytic activity): 0.29 mM ARS, 0.31 mM PNS, 2 mM XO, 0.15 mM MB, 1.5 mM MG, and 1.25 mM MO. During the kinetic experiments, wide ranges of substrate concentrations were studied (0.01-2 mM for the dyes, 0.1-40 mM for hydrogen peroxide).

The reaction mixture was then stirred at 25°C in the dark, and the reaction was monitored photometrically using the $\lambda_{max}$ and molar extinction coefficients, reported in Table 1.

The actual hydrogen peroxide concentration was determined by $KMnO_4$ titration.

When the reaction pH was not 7, the pH was corrected to 7 with 0.1 mL of potassium phosphate buffer 1 M (pH 7) before an absorbance measurement. UV–vis spectra in the range 230–700 nm were also recorded. Different McIlvaine buffers were used in the pH range 3–8.

When used, redox mediators (4-hydroxy-2,2,6,6-tetramethylpiperidine-1-oxyl OH- TEMPO; *N*-hydroxyphthalimide NPH; *N*-hydroxysuccinimide NHS; and *N*-hydroxybenzotriazole NHT) were added to the reaction medium at a final concentration of 1 mM.

For the MnP-like assays, all the experiments were repeated as above in the presence of 1 mM $MnSO_4$ and 50 mM malonic acid. During multicyclic runs, the catalysts were regenerated between consecutive cycles by repeated washings with water and 2-propanol and by drying at 50°C.

Total organic carbon determination was performed by automatic analyzer TOC 5000A (Shimadzu), equipped with Pt-$Al_2O_3$ oxidizing catalyst. The volume of injection was 25 μL. The calibration curve was obtained using a range of standard solutions of potassium phthalate (0–200 ppm).

Michaelis-Menten kinetic parameters were calculated by using R 2.5.1 software (*R Foundation for Statistical Computing*, Vienna).

## Enzymatic Assays

When horseradish peroxidase (HRP) was used, up to 20 E.U. were present in a final volume of 1 mL of 25 mM buffer (pH 4, 5, 6, or 7) with 8.8 mM $H_2O_2$. Due to significant substrate inhibition, lower dyes concentrations were used: 119 μM ARS; 31 μM PNS; 200 μM XO; 15 μM MB; 150 μM MG; 65 μM MO.

The same conditions were adopted for the experiments with laccase (23 E.U. added); every experiment was performed in the presence and absence of $H_2O_2$.

In the case of lignin peroxidase (LiP), 0.2 E.U. were present in the final volume of 1 mL, with 0.176 mM $H_2O_2$ due to strong substrate inhibition.

Moreover, to mimic manganese peroxidase (MnP) action, other experiments were conducted using Mn(III) as the putative oxidizing agent. One mM manganese triacetate was dissolved in 50 mM sodium malonate buffer, pH 6.5 [27], and the final mixture contained this solution along with dyes at the same concentrations, as mentioned above.

# CONCLUSION

IPS/MnTSPP has proved able to efficiently degrade all tested dyes, regardless of their chemical class, under very mild conditions: standard pressure, room temperature and neutral pH, using no organic solvents, and the most environmental-friendly oxidant. The yields and time requirements of the process were quite promising for a potential industrial application. Furthermore, a multi-cycle capability was described, even if only 30% of catalytic activity is retained for 8–10 cycles.

In comparison to enzymatic systems, the biomimetic catalyst was more efficient and had wider substrate specificity. Therefore, after proper optimization of the process, this catalyst is worth consideration as a possible medium/large scale application in textile wastewater treatment.

## ACKNOWLEDGMENTS

This research was partly funded by Fondazione Banco di Sardegna (Project Code 861/2012) and Regione Autonoma Sardegna (PO Sardegna FSE 2007–2013, L.R.7/2007, Bando Giovani Ricercatori, Project CRP1_27, and PO Sardegna FSE 2007–2013, L.R.7/2007, Ricerca di Base Bando 2008, Project CRP1_408).

## AUTHORS' CONTRIBUTIONS

PZ performed the main part of the experiments and drafted the manuscript. MP performed TOC and enzymatic assays. ACR, and AR helped analyzing the data and draft the manuscript. ES conceived the study, coordinated the study, and worked on data analysis and interpretation. All the authors read and approved the final manuscript.

## REFERENCES

1. Forgacs E, Cserhati T, Oros G: Removal of synthetic dyes from wastewaters: a review. Environ Int. 2004, 30: 953-971. 10.1016/j.envint.2004.02.001.
2. Azizi A, Moghaddam MRA, Arami M: Application of wood waste for removal of reactive blue 19 from aqueous solutions: optimization through response surface methodology. Environ Eng Manag J. 2012, 11: 795-804.
3. Brînzilă CI, Ciobanu R, Brillas E: Effect of experimental parameters on crystal violet mineralization by electro-fenton processes. Environ Eng Manag J. 2012, 11: 517-523.

4. Tsuda S, Murakami M, Matsusaka N, Kano K, Taniguchi K, Sasaki YF: DNA Damage induced by red food dyes orally administered to pregnant and male mice. Toxicol Sci. 2001, 61: 92-99. 10.1093/toxsci/61.1.92.
5. Walthall WK, Stark JD: The acute and chronic toxicity of two xanthene dyes, fluorescein sodium salt and phloxine B, to Daphnia pulex. Environ Pollut. 1999, 104: 207-215. 10.1016/S0269-7491(98)00189-4.
6. Pierce J: Colour in textile effluents - the origins of the problem. J Soc Dye Colour. 1994, 110: 131-133.
7. Colar LA, Ilinoiu EC, Manea F, Pode R, Lazau C: Photocatalytic activity of a silver doped TIO 2 modified zeolite in the degradation of reactive yellow 125 azo dye. Environ Eng Manag J. 2012, 11: 61-68.
8. Zucca P, Rescigno A, Olianas A, Maccioni S, Sollai F, Sanjust E: Induction, purification, and characterization of a laccase isozyme from Pleurotus sajor-caju and the potential in decolorization of textile dyes. J Mol Catal B Enzym. 2011, 68: 216-222. 10.1016/j.molcatb.2010.11.008.
9. Pakshirajan K, Jaiswal S, Das RK: Biodecolourization of azo dyes using phanerochaete chrysosporium: effect of culture conditions and enzyme activities. J Sci Ind Res. 2011, 70: 987-991.
10. Córdoba A, Magario I, Ferreira ML: Experimental design and MM2-PM6 molecular modelling of hematin as a peroxidase-like catalyst in Alizarin Red S degradation. J Mol Catal A Chem. 2012, 355: 44-60.
11. Zhu L, Ghosh T, Park CY, Meng ZD, Oh WC: Enhanced sonocatalytic degradation of rhodamine B by graphene-TiO 2 composites synthesized by an ultrasonic-assisted method. Chin J Catal. 2012, 33: 1276-1283. 10.1016/S1872-2067(11)60430-0.
12. Oh WC, Chen M, Cho K, Kim C, Meng Z, Zhu L: Synthesis of graphene-CdSe composite by a simple hydrothermal method and its photocatalytic degradation of organic dyes. Chin J Catal. 2011, 32: 1577-1583. 10.1016/S1872-2067(10)60264-1.
13. Asgher M, Kamal S, Iqbal HMN: Improvement of catalytic efficiency, thermo-stability and Dye decolorization capability

of pleurotus ostreatus IBL-02 laccase by hydrophobic Sol Gel entrapment. Chem Centr J. 2012, 110:

14. Stengl V, Matys Grygar T, Henych J, Kormunda M: Hydrogen peroxide route to Sn-doped titania photocatalysts. Chem Centr J. 2012, 113:
15. Velmurugan R, Sreedhar B, Swaminathan M: Nanostructured AgBr loaded TiO 2: an efficient sunlight active photocatalyst for degradation of reactive Red 120. Chem Centr J. 2011, 5:
16. Chiavola A: Textiles. Water Environ Res. 2012, 84: 1511-1532. 10. 2175/106143012X13407275695391.
17. Zucca P, Rescigno A, Sanjust E: Ligninolytic peroxidase-like activity of a synthetic metalloporphine immobilized onto mercapto-grafted crosslinked PVA inspired by the active site of cytochrome P450. Chin J Catal. 2011, 32: 1663-1666. 10.1016/S1872-2067(10)60277-X.
18. Faria AL, Mac Leod TOC, Barros VP, Assis MD: Hydrocarbon oxidation catalyzed by iron and manganese porphyrins anchored on aminofunctionalized supports. J Brazil Chem Soc. 2009, 20: 895-906. 10.1590/S0103-50532009000500014.
19. Dos Santos JS, Faria AL, Da Silva Amorin PM, La Luna FM, Caiado KL, Silva DOC E, Sartoratto PPC, Assis MD: Iron(III) porphyrin covalently supported onto magnetic amino-functionalized nanospheres as catalyst for hydrocarbon and herbicide oxidations. J Brazil Chem Soc. 2012, 23: 1411-1420. 10.1590/S0103-50532012005000005.
20. Díaz-Díaz G, Blanco-López MC, Lobo-Castañón MJ, Miranda-Ordieres AJ, Tuñón-Blanco P: Hemo-acrylic polymers as catalyst for the oxidative dehalogenation of 2,4,6-trichlorophenol. Chloroperoxidase's mimic imprinting effects. J Mol Catal A Chem. 2012, 353–354: 117-121.
21. Tian P, Gao B, Chen Y: Synchronous synthesis and immobilization of pyridyl porphyrins on crosslinked polystyrene microspheres and catalytic oxidation performance of immobilized cationic Co-porphyrins. Chin J Catal. 2011, 32: 483-489.
22. Zucca P, Vinci C, Sollai F, Rescigno A, Sanjust E: Degradation of Alizarin Red S under mild experimental conditions by immobilized 5,10,15,20-tetrakis(4-sulfonatophenyl)porphine-Mn(III) as a

biomimetic peroxidase-like catalyst. J Mol Catal A Chem. 2008, 288: 97-102. 10.1016/j.molcata.2008.04.001.

23. Zucca P, Vinci C, Rescigno A, Dumitriu E, Sanjust E: Is the bleaching of phenosafranine by hydrogen peroxide oxidation catalyzed by silica-supported 5,10,15,20-tetrakis-(sulfonatophenyl)porphine-Mn(III) really biomimetic?. J Mol Catal A Chem. 2010, 321: 27-33. 10.1016/j.molcata.2010.01.015.
24. Zucca P, Mocci G, Rescigno A, Sanjust E: 5,10,15,20-Tetrakis(4-sulfonato-phenyl)porphine-Mn(III) immobilized on imidazole activacted silica as a novel lignin-peroxidase-like biomimetic catalyst. J Mol Catal A Chem. 2007, 278: 220-227. 10.1016/j.molcata.2007.09.017.
25. Zucca P, Sollai F, Garau A, Rescigno A, Sanjust E: Fe(III)-5,10,15,20-Tetrakis(pentafluorophenyl)porphine supported on pyridyl-functionalized, crosslinked poly(vinylalcohol) as a biomimetic versatile-peroxidase-like catalyst. J Mol Catal A Chem. 2009, 306: 89-96. 10.1016/j.molcata.2009.02.029.
26. Cojocariu AM, Mutin PH, Dumitriu E, Vioux A, Fajula F, Hulea V: Removal of dimethylsulfoxide from wastewater using mild oxidation with H2O2 over Ti-based catalysts. Chemosphere. 2009, 77: 1065-1068. 10.1016/j.chemosphere.2009.08.059.
27. Wariishi H, Valli K, Gold MH: Manganese(II) oxidation by manganese peroxidase from the basidiomycete phanerochaete-chrysosporium - kinetic mechanism and role of chelators. J Biol Chem. 1992, 267: 23688-23695.
28. Rescigno A, Zucca P, Flurkey A, Inlow J, Flurkey WH: Identification and discrimination between some contaminant enzyme activities in commercial preparations of mushroom tyrosinase. Enzyme Microb Technol. 2007, 41: 620-627. 10.1016/j.enzmictec.2007.05.009.
29. Sollai F, Zucca P, Sanjust E, Steri D, Rescigno A: Umbelliferone and esculetin: inhibitors or substrates for polyphenol oxidases?. Biol Pharm Bull. 2008, 31: 2187-2193. 10.1248/bpb.31.2187.
30. Zucca P, Littarru M, Rescigno A, Sanjust E: Cofactor recycling for selective enzymatic biotrasformation of cinnamaldehyde to cinnamyl alcohol. Biosci Biotechnol Biochem. 2009, 73: 1224-1226. 10.1271/bbb.90025.

31. Rescigno A, Casañola-Martin GM, Sanjust E, Zucca P, Marrero-Ponce Y: Vanilloid derivatives as tyrosinase inhibitors driven by virtual screening-based QSAR models. Drug Test Anal. 2011, 3: 176-181. 10.1002/dta.187.

32. Sanjust E, Cocco D, Curreli N, Rescigno A, Sollai F, Bannister JV: Flavin-grafted poly(vinyl alcohol): preparation and properties. J Appl Polym Sci. 2002, 85: 2471-2477. 10.1002/app.10877.

CHAPTER 7

# FUSING A CARBOHYDRATE-BINDING MODULE INTO THE ASPERGILLUS USAMII Β-MANNANASE TO IMPROVE ITS THERMOSTABILITY AND CELLULOSE-BINDING CAPACITY BY IN SILICO DESIGN

**Cun-Duo Tang[1], Jian-Fang Li[2], Xi-Huan Wei[2], Rou Min[3], Shu-Juan Gao[3], Jun-Qing Wang[1], Xin Yin[1] , Min-Chen Wu[4]**

[1]School of Biotechnology and Key Laboratory of Industrial Biotechnology, Ministry of Education, Jiangnan University, Wuxi, Jiangsu, China

[2]School of Food Science and Technology, Jiangnan University, Wuxi, Jiangsu, China

[3]School of Pharmaceutical Science, Jiangnan University, Wuxi, Jiangsu, China

[4]Wuxi Medical School, Jiangnan University, Wuxi, Jiangsu, China

## ABSTRACT

The AuMan5A, an acidophilic glycoside hydrolase (GH) family 5 β-mannanase derived from *Aspergillus usamii* YL-01-78, consists of an only catalytic domain (CD). To perfect enzymatic properties of the AuMan5A, a family 1 carbohydrate-binding module (CBM) of the

*Trichoderma reesei* cellobiohydrolase I (TrCBH I), having the lowest binding free energy with cellobiose, was selected by *in silico* design, and fused into its C-terminus forming a fusion β-mannanase, designated as AuMan5A-CBM. Then, its encoding gene, *Auman5A-cbm*, was constructed as it was designed theoretically, and expressed in *Pichia pastoris* GS115. SDS-PAGE analysis displayed that both recombinant AuMan5A-CBM (reAuMan5A-CBM) and AuMan5A (reAuMan5A) were secreted into the cultured media with apparent molecular masses of 57.3 and 49.8 kDa, respectively. The temperature optimum of the reAuMan5A-CBM was 75°C, being 5°C higher than that of the reAuMan5A. They were stable at temperatures of 68 and 60°C, respectively. Compared with reAuMan5A, the reAuMan5A-CBM showed an obvious decrease in $K_m$ and a slight alteration in $V_{max}$. In addition, the fusion of a CBM of the TrCBH I into the AuMan5A contributed to its cellulose-binding capacity.

# INTRODUCTION

β-Mannanases (EC 3.2.1.78), abbreviated from β-1,4-D-mannan mannohydrolases, can hydrolyze the internal β-1,4-D-mannosidic linkages of mannans. They could be applied in industrial processes, such as bleaching pulps, depolymerizing anti-nutritional factors in feedstuffs, the production of mannooligosaccharides, and extracting oils from leguminous seeds [1]. To date, many researches have been performed on exploiting novel β-mannanases with good properties, improving their catalytic activities by mutating enzyme-producing strains and optimizing fermentation conditions, as well as producing β-mannanases on an industrial scale [2], [3]. However, the applicability of β-mannanases, exemplified by preparing feedstuffs and bleaching pulps, was limited by their low stability at the high temperature and/or extreme pH. To meet the increasing needs for β-mannanases, more interests are being focused on modifying their molecular structures by means of genetic engineering [4].

Based on the sequence alignment, 25,580 CBMs in the CAZY database (July, 2012) were classified into 64 families. The recognized function of CBMs was to bring enzymes into close vicinity to their substrates by binding carbohydrates [5]. Some studies also demonstrated that the fusion of CBMs into enzymes improved the catalytic activity and/or thermostability. For instances, delignification efficiency of the

*Pycnoporus cinnabarinus* laccase for softwood kraft pulp bleaching was improved by fusing a family 1 CBM of the *A. niger* cellobiohydrolase B into the laccase [6]. After two CBMs of the *Thermobifida fusca* and *Cellulomonas fimi* cellulases were fused into the *T. fusca* cutinase, respectively, both fusion cutinases exhibited a dramatic increase of up to 3-fold in the amount of fatty acids released from cotton fiber [7].

Almost all β-mannanases reported have been classified into GH families 5, 26 and 113 (http://www.cazy.org/fam/acc_GH.html) [8]. The family 5 β-mannanases either contain an only CD [9] or, besides a CD, carry an additional CBM located at the C-terminus such as *T. reesei*β-mannanase [10] or at the N-terminus such as *Phanerochaete chrysosporium* β-mannanase [11]. In our previous work, an AuMan5A-encoding gene (*Auman5A*) was cloned and analyzed. The multiple sequence alignment among family 5 β-mannanases displayed that no CBM was found in the AuMan5A [12]. To perfect the AuMan5A's properties, our present work designed an AuMan5A-CBM by fusing a CBM of the TrCBH I into the AuMan5A. Then, the *Auman5A-cbm* was constructed as it was designed theoretically, and expressed in *P. pastoris* GS115. Moreover, enzymatic properties of the reAuMan5A-CBM and reAuMan5A were analyzed and compared.

# MATERIALS AND METHODS

## Strains, Vectors and Media

*A. usamii* YL-01-78 and *T. reesei* LW-22, isolated from the soil in China, were used as donors of the *Auman5A* and CBH I-encoding gene (*Trcbh*), respectively. *E. coli* JM109 and pUCm-T (Sangon, Shanghai, China) were used for gene cloning and DNA sequencing. Two recombinant T-vectors, pUCm-T-*Auman5A* and pUCm-T-*Trcbh*, were constructed according to the sequences of the *Auman5A* (GenBank accession: HQ839639) and *Trcbh* (GL985084). *E. coli*DH5α and pPIC9K (Invitrogen, San Diego, CA) were used for the construction of recombinant expression vectors. *E. coli* JM109 and DH5α were cultured in a LB medium. *P. pastoris* GS115 was cultured in a yeast extract peptone dextrose (YPD) medium, and its transformant cultured and induced in following media that were prepared as described in the manual of Multi-Copy *Pichia* Expression Kit

(Invitrogen): minimal dextrose (MD), geneticin G418-containing YPD, buffered glycerol-complex (BMGY) and buffered methanol-complex (BMMY).

## Bioinformatics Analysis of the β-mannanase and CBM Sequences

The multiple protein sequence alignment of the CBMs was performed using the ClustalW2 program (http://www.ebi.ac.uk/Tools/msa/clustalw2/). The putative N-glycosylation site of the β-mannanase was located using the NetNGlyc program (http://www.cbs.dtu.dk/services/NetNGlyc/). The Protparam program (http://au.expasy.org/tools/protparam.html) was used for predicting the β-mannanase physicochemical properties. The phylogenetic tree of the CBMs was constructed using both the ClustalW2 program and MEGA 4.0 software. The three-dimensional (3-D) structures of the CBMs were predicted by homology modeling using the MODELLER 9.9 program (http://salilab.org/modeller/).

## *In silico* Design of the Fusion β-mannanase

The candidate CBMs were chosen from the phylogenetic tree, and their 3-D structures were modeled based on a CBM crystal one of the TrCBH I (PDB code: 1CBH). While the 3-D structural information of cellobiose, used as the ligand, was handled using the GlycoBioChem PRODRG (http://davapc1.bioch.dundee.ac.uk/prodrg/submit.html). Then, the interaction between the candidate CBM and cellobiose was predicted by molecular docking (MD) simulation. The binding free energy of the CBM with cellobiose, contrary to their affinity [13], was calculated using the AutoDock 4.2 program (http://autodock.scripps.edu), which combines a rapid binding free energy evaluation through precalculated grids of affinity potentials with a variety of search algorithms to find the most suitable binding position for a ligand on a given macromolecule [14]. Finally, a CBM having the lowest binding free energy was selected, and fused into the AuMan5A forming an AuMan5A-CBM.

## Construction of the Fusion β-mannanase Gene

The *Auman5A-cbm* was constructed by fusing the 3′-end region (*lcbm*) of the *Trcbh*, which encodes both a Ser/Thr/Pro-rich linker and a CBM, into 3′-end of the *Auman5A* by overlapping PCR. The *Auman5A* was amplified from the pUCm-T-*Auman5A* with primers F1 (5′-GAATTCTCCTTCGCCAG CACCTC-3′ with an *Eco*R I site, underlined) and R1 (5′-GAGGGTTGCCGGCACTATC AATAGCAGC-3′). The *lcbm* was amplified from the pUCm-T-*Trcbh* with primers F2 (5′-TGATAGTGCCGGC AACCCTCCCGGCG-3′) and R2 (5′-GCGGCCGCTTACAGGCACTGAGAGTAG-3′, with a *Not* I site, underlined). Using the *Auman5A* and *lcbm* as primers and templates, the first-round overlapping PCR for the *Auman5A-cbm* was performed as follows: a denaturation at 94°C for 4 min; 10 cycles of at 94°C for 30 s, 52°C for 30 s, 72°C for 75 s; an elongation at 72°C for 10 min. Then, F1 and R2 were added to the above PCR system to run the second-round PCR under the same conditions, except 30 cycles and an annealing temperature of 56°C. The amplified target band was cloned into pUCm-T, and transformed into *E. coli* JM109. The recombinant T-vector, pUCm-T-*Auman5A-cbm*, was confirmed by DNA sequencing.

## Transformation of the Recombinant Expression Vectors

The *Auman5A-cbm* and *Auman5A* were excised from the pUCm-T-*Auman5A-cbm* and pUCm-T-*Auman5A* by digestion with *Eco*R I and *Not* I, and inserted into pPIC9K, followed by transforming them into *E. coli* DH5α, respectively. Recombinant expression vectors, pPIC9K-*Auman5A-cbm* and pPIC9K-*Auman5A*, were confirmed by sequencing. The resulting recombinant vectors were separately linearized with *Sac* I, and transformed into *P. pastoris*GS115 using a Gene Pulser Apparatus (Bio-Rad, Hercules, CA).

## Screening and Expression of the *P. pastoris* Transformants

All *P. pastoris* transformants were primarily screened based on their ability to grow on a MD plate, then inoculated successively on G418-containing YPD plates at increasing concentrations of 1.0, 2.0 and 4.0 mg/mL to screen multiple copies of the *Auman5A-cbm* and *Auman5A*, respectively. *P. pastoris* GS115 transformed with pPIC9K vector without any insert was used as the negative control. Expression of the gene *Auman5A-cbm* or *Auman5A* in *P. pastoris* GS115 was performed according to the instruction of Multi-Copy Pichia Expression Kit (Invitrogen) with some modifications [15].

## Enzyme Activity and Protein Assays

β-Mannanase activity was determined by measuring the amount of reducing sugars from locust bean gum (Sigma, St. Louis, MA), using the 3,5-dinitrosalicylic acid (DNS) method as reported previously [15]. One unit (U) of β-mannanase activity was defined as the amount of enzyme liberating 1.0 $\mu$mol of reducing sugar equivalent per min under the standard assay conditions (at pH 3.6 and 60°C for 10 min). The protein concentration was measured with the BCA-200 Protein Assay Kit (Pierce, Rockford, IL). Sodium dodecyl sulfate-polyacrylamide gel electrophoresis (SDS-PAGE) was carried out on a 12.5% gel using the reported method [16], and the isolated proteins were visualized by staining with Coomassie Brilliant Blue R-250 (Sigma).

## Purification of the Recombinant β-mannanases

After 96-h induction, the cultured medium of the transformant was centrifuged to remove cells. A total of 60 mL of supernatant was brought to 75% saturation by adding solid ammonium sulfate. The resulting precipitate was collected, dissolved in 6.0 mL of 20 mM $Na_2HPO_4$-$NaH_2PO_4$ buffer (pH 7.0), and dialyzed against the same buffer overnight. The dialyzed solution was concentrated to 1.5 mL by ultrafiltration using a 30-kDa cutoff membrane (Millipore, Billerica, MA), and then loaded onto a Sephadex G-75 column (Amersham Pharmacia Biotech, Sweden; 1.6 × 80 cm), followed by elution with the same buffer at a flow rate of 0.3 mL/min. Aliquots of 1.5 mL eluent only containing the reAuMan5A-CBM or reAuMan5A were pooled and concentrated for further studies.

## Temperature Optimum and Stability Assays

The temperature optima of the reAuMan5A-CBM and reAuMan5A were determined under the standard assay conditions, except temperatures ranging from 50 to 80°C. To estimate the thermostability, the reAuMan5A-CBM or reAuMan5A was incubated at pH 3.6 and various temperatures (45–75°C) for 1.0 h, then the residual enzyme activity was measured under the standard assay conditions. The thermostability here was defined as a temperature, at or below which the residual reAuMan5A-CBM or reAuMan5A activity retained more than 85% of its original activity.

## Enzymatic Kinetic Parameter Assays

Reaction rate (U/mg) of the reAuMan5A-CBM or reAuMan5A was measured under the standard assay conditions, except locust bean gum concentrations ranging from 1.0 to 10 mg/mL. The reaction rate versus the substrate concentration was plotted to confirm whether the catalyzing mode of the reAuMan5A-CBM or reAuMan5A conforms to the Michaelis-Menten equation. The enzymatic kinetic parameters, $K_m$ and $V_{max}$ values, were graphically determined from the Lineweaver-Burk plotting.

## Cellulose-binding Capacity Assay

The cellulose-binding test was performed according to the method [17] with some modifications. The cultured supernatant of the transformant was dialyzed against the 20 mM $Na_2HPO_4$-$NaH_2PO_4$ buffer (pH 7.0), and loaded onto a glass column (0.8 × 10 cm), where 1.0 g crystalline cellulose Avicel PH-101 (Sigma) was filled, at a flow rate of 0.2 mL/min, followed by washing with the same buffer. Total volume of the effluent was scaled. The binding capacity of the reAuMan5A-CBM or reAuMan5A was estimated by measuring the activity of unbound β-mannanase in the effluent.

# RESULTS

## *In silico* Design of the Fusion β-mannanase

From the CAZY database, we randomly selected 23 family 1 CBM sequences. Then, the CBM phylogenetic tree was constructed (Fig. 1). Based on the topology of the tree, those CBMs were further grouped into 4 subfamilies. One to three CBMs from each subfamily were chosen as the candidate CBMs. The interaction between 3-D structures of the candidate CBM and cellobiose was predicted by MD simulation, and then the binding free energy was calculated using the AutoDock 4.2 program (Table 1). As a result, a CBM of the TrCBH I having the lowest binding free energy of −2.27 kcal/mol was selected, and fused into the C-terminus of AuMan5A. The molecule-docked conformation of the CBM with cellobiose was illustrated (Fig. 2).

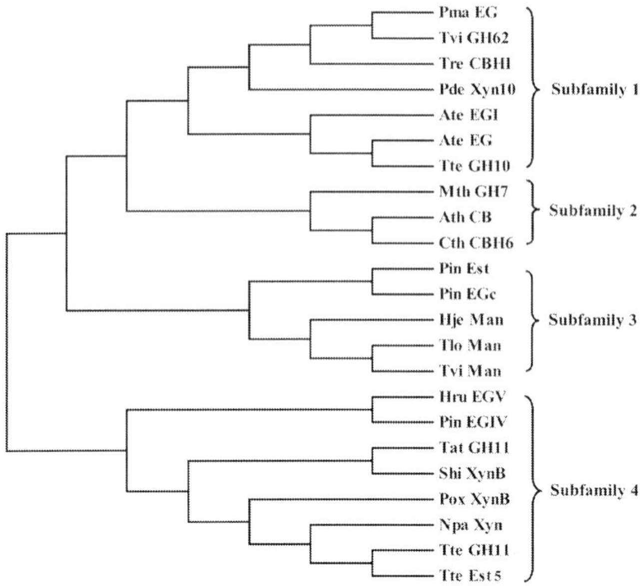


**Figure 1:** The phylogenetic tree showing the evolutionary relativity and homology among family 1 CBMs.

The CBM-containing hydrolases are as follows: Pma EG, *P. marneffei* endoglucanase (XP_002152969); Tvi GH62, *T. virens* GH family 62 enzyme (EHK19840); Tre CBH I, *T. reesei* cellobiohydrolase I (EGR44817); Pde Xyn10, *P. decumbens* family 10 xylanase (ADX86896); Ate EG I, *A. terreus* endoglucanase I (XP_001217291); Ate EG, *A. terreus*endoglucanase (AAW68436); Tte GH10, *T. terrestris* GH family 10 enzyme (XP_003653793); Mth GH7, *M. thermophila* GH family 7 enzyme (XP_003663441); Ath CB, *A. thermophilum* cellobiosidase (CAM98445); Cth CBH6, *C. thermophilum* family 6 cellobiohydrolase (AAY88915); Pin Est, *P. indica* acetylxylan esterase (CCA73570); Pin EGc, *P. indica* endoglucanase c (CCA67649); Hje Man, *H. jecorina* mannanase (AAA34208); Tlo Man, *T. longibrachiatum* mannanase (ADN93457); Tvi Man, *T. viride*mannanase (AFP95336); Hru EG V, *H. rufa* endoglucanase V (AAQ21385); Pin EG IV, *P. indica* endoglucanase IV (CCA70703); Tat GH11, *T. atroviride* GH family 11 enzyme (EHK46770); Shi XynB, *S. hirsutum* xylanase B (EIM91441); Pox XynB, *P. oxalicum*xylanase B (ADV31286); Npa Xyn, *N. patriciarum* xylanase (ABW04217); Tte GH11, *T. terrestris* GH family 11 enzyme (XP_003649436); Tte Est5, *T. terrestris* family 5 carbohydrate esterase (XP_003653797).

https://doi.org/10.1371/journal.pone.0064766.g001

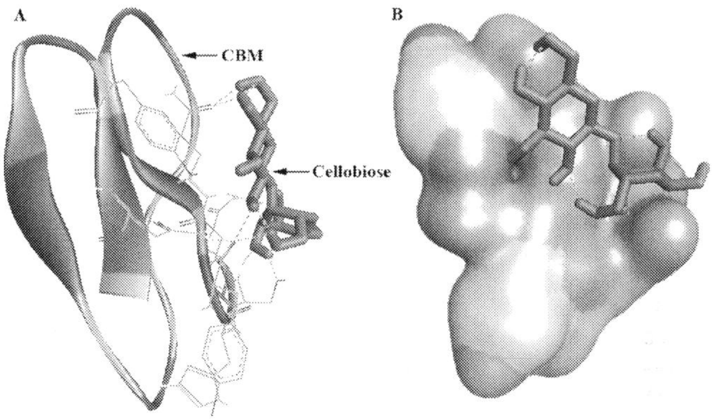


**Figure 2:** The molecule-docked conformation between 3-D structures of a CBM of the TrCBH I and cellobiose.

(A) a cartoon model and (B) a surface model.

https://doi.org/10.1371/journal.pone.0064766.g002

**Table 1:** The binding free energy of the candidate CBM combining with cellobiose

https://doi.org/10.1371/journal.pone.0064766.t001

| CBM-containing hydrolase | GenBank accession no. | Length of CBM (aa) | Aromatic residues in the groove surface | Apolar area/energy of CBM($Å^2$) | Binding free energy (kcal/mol) |
|---|---|---|---|---|---|
| Pma EG | XP_002152969 | 34 | $Trp^5$, $Tyr^{29}$, $Tyr^{30}$ | 1775.44 | −1.56 |
| Tre CBH I | EGR44817 | 36 | $Tyr^5$, $Tyr^{13}$, $Tyr^{31}$, $Tyr^{32}$ | 1644.23 | **−2.27** |
| Ate EG | AAW68436 | 33 | $Trp^{10}$, $Tyr^{20}$, $Trp^{28}$, $Tyr^{29}$ | 1780.38 | −1.21 |
| Pin Est | CCA73570 | 33 | $Tyr^2$, $Tyr^{10}$, $Tyr^{23}$, $Trp^{27}$, $Tyr^{28}$ | 1753.77 | −1.64 |
| Tvi Man | AFP95336 | 34 | $Tyr^2$, $Tyr^8$, $Tyr^{23}$, $Tyr^{26}$, $Trp^{27}$, $Tyr^{28}$ | 1776.76 | −1.98 |
| Hru EG V | AAQ21385 | 33 | $Tyr^2$, $Trp^{10}$, $Trp^{28}$, $Tyr^{29}$ | 1727.44 | −1.65 |
| Pox XynB | ADV31286 | 20 | $Trp^{15}$, $Tyr^{16}$ | 1275.21 | −1.78 |
| Tte GH11 | XP_003649436 | 33 | $Trp^3$, $Trp^{11}$, $Trp^{28}$, $Tyr^{29}$ | 1725.45 | −1.16 |

The apolar area/energy of CBM were calculated by the GETAREA servicer (http://curie.utmb.edu/getarea.html).
doi:10.1371/journal.pone.0064766.t001

## Construction of the Fusion β-mannanase Gene

About 1050- and 210-bp bands of the *Auman5A* and *lcbm* were amplified from the pUCm-T-*Auman5A* and pUCm-T-*Trcbh*, respectively. Then, they were subjected to the overlapping PCR to construct the fusion gene. As a result, an about 1250-bp band of the *Auman5A-cbm* was amplified and inserted into pUCm-T. DNA sequencing result verified that the cloned *Auman5A-cbm* is 1241 bp in length (containing *Eco*R I and *Not* I sites), encoding an AuMan5A-CBM of 408 amino acid residues with a theoretical molecular mass of 43,798 Da and a pI of 4.22. The AuMan5A-CBM consists of a 345-aa CD from the AuMan5A, and both a 27-aa linker and a 36-aa CBM from the TrCBH I (Fig. 3).

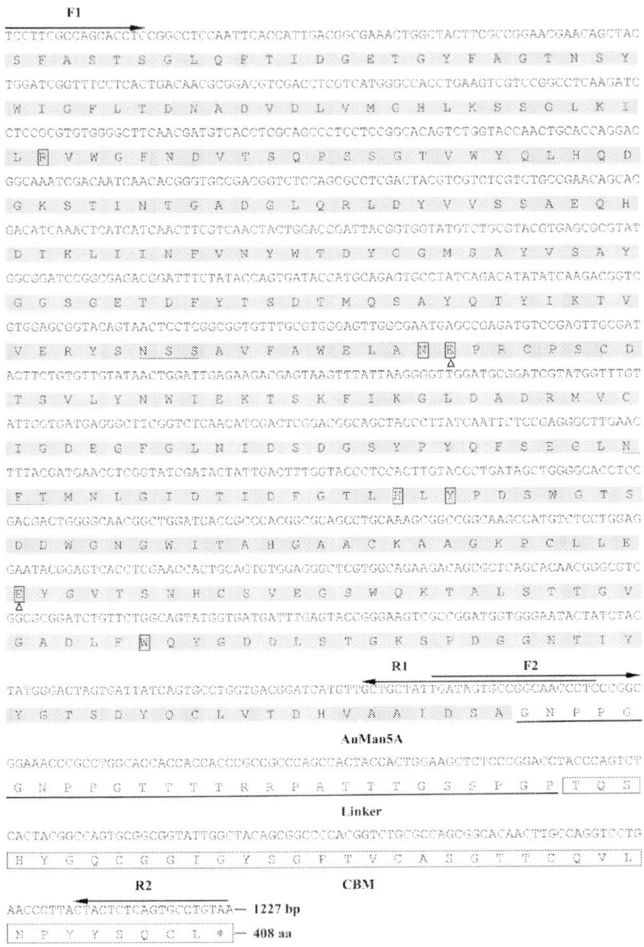


**Figure 3:** Nucleotide sequence of the *Auman5A-cbm* **and its deduced amino acid sequence of the AuMan5A-CBM.**

The amino acid residues of the AuMan5A are marked in grayed background. A linker of the *T. reesei* CBH I is underlined and its CBM is boxed. Two triangles below the boxed letters indicate the catalytic residues

(E168 and E276) and five active site residues (R52, N167, H241, Y243, and W306) are located in grayed boxes. Two putative N-glycosylation sites of the AuMan5A-CBM are underlined in grayed background. The bold arrows above the letters represent the PCR primers.

https://doi.org/10.1371/journal.pone.0064766.g003

## Screening and Expression of the *P. pastoris* Transformants

A total of 30 *P. pastoris* transformants with *Auman5A* respectively resistant to 1.0, 2.0 and 4.0 mg/mL of geneticin G418, numbered as *P. pastoris* GSAuM1-1 to GSAuM1-10, GSAuM2-1 to GSAuM2-10 and GSAuM4-1 to GSAuM4-10, and 30 transformants with *Auman5A-cbm*, numbered as *P. pastoris* GSAuMC1-1 to GSAuMC1-10, GSAuMC2-1 to GSAuMC2-10 and GSAuMC4-1 to GSAuMC4-10, were picked out for flask expression tests. *P. pastoris* GS115 transformed with pPIC9K, numbered as *P. pastoris* GSC, was used as the negative control. After induction by adding 1.0% (v/v) methanol at 24 h intervals for 96 h, the cultured supernatants were used for β-mannanase activity assay. As a result, two transformants with the highest reAuMan5A-CBM and reAuMan5A activities of 40.6 and 44.1 U/mL, numbered as *P. pastoris* GSAuMC4-5 and GSAuM4-8, were selected, respectively. No β-mannanase activity of *P. pastoris* GSC was detected under the same expression conditions.

## Purification of the Recombinant β-mannanases

The amount of the reAuMan5A-CBM or reAuMan5A in the cultured supernatant of the GSAuMC4-5 or GSAuM4-8 accounted for more than 85% of that of the total protein. So they were purified to homogeneity only by a simple combination of ammonium sulfate precipitation, ultrafiltration and Sephadex G-75 gel filtration (Fig. 4). Specific activities of the purified reAuMan5A-CBM and reAuMan5A, towards locust bean gum under the standard assay conditions, were 312.3 and 341.3 U/mg, respectively.

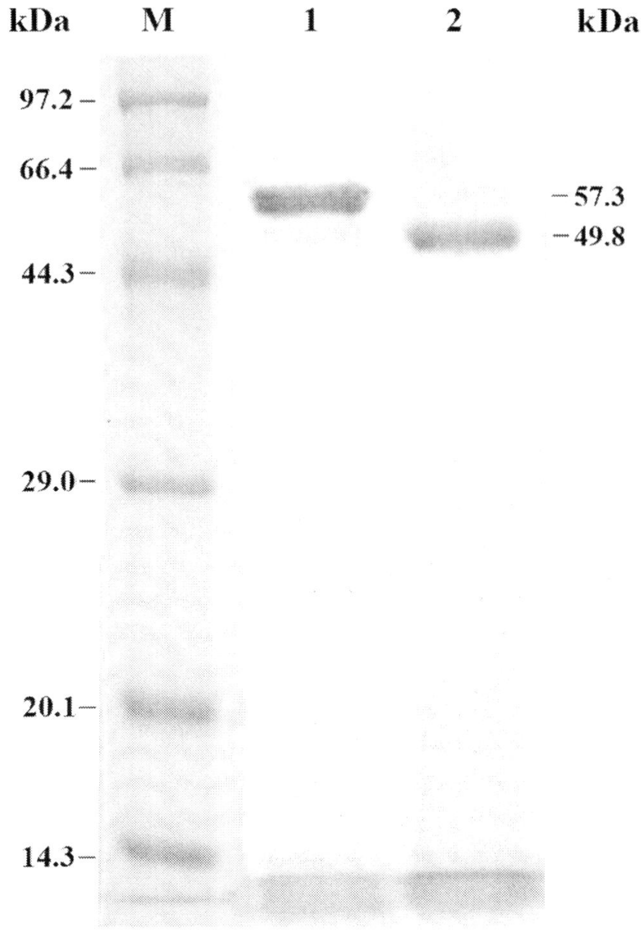


**Figure 4:** SDS-PAGE analysis of the purified reAuMan5A-CBM and reAu-Man5A.

Lane M, standard protein marker; lane 1, the AuMan5A-CBM with an apparent molecular mass of 57.3 kDa; lane 2, the reAuMan5A with an apparent molecular mass of 49.8 kDa.

https://doi.org/10.1371/journal.pone.0064766.g004

SDS-PAGE analysis of the purified reAuMan5A-CBM and reAuMan5A showed two single protein bands with apparent molecular masses of 57.3 and 49.8 kDa, respectively (Fig. 4), which were much larger than their respective theoretical ones (43,798 and 37,614 Da).

*P. pastoris* enables some post-translational modifications, including the assembly of disulfide bond, exclusion of the signal peptide, and N-glycosylation, etc. Analysis showed that there are two putative N-glycosylation sites (N156-S157-S158 and N225-F226-T227) in the AuMan5A-CBM or AuMan5A sequence (Fig. 3). To verify whether the difference between apparent and theoretical molecular masses was resulted from N-glycosylation, The N-deglycosylation assay was performed using an endoglycosidase H (New England Biolabs, Ipswich, MA) according to the manufacturer's instruction. As a result, molecular sizes of the N-deglycosylated reAuMan5A-CBM and reAuMan5A on SDS-PAGE (data not shown) were good agreement with their respective theoretical ones. So the larger apparent molecular masses may be caused by N-glycosylation. Carbohydrate contents of the purified reAuMan5A-CBM and reAuMan5A were determined to be 19.8 and 21.3%, respectively, using the phenol-sulfuric acid method [18].

## Enzymatic Properties of the Recombinant Enzymes

The temperature optima of the reAuMan5A-CBM and reAuMan5A were 75 and 70°C (measured at pH 3.6), respectively (Fig. 5A). They were highly stable at temperatures of 68 and 60°C, respectively, but over which their residual activities declined rapidly and only retained 62.3 and 36.8% of the original ones at 75°C, respectively (Fig. 5B).

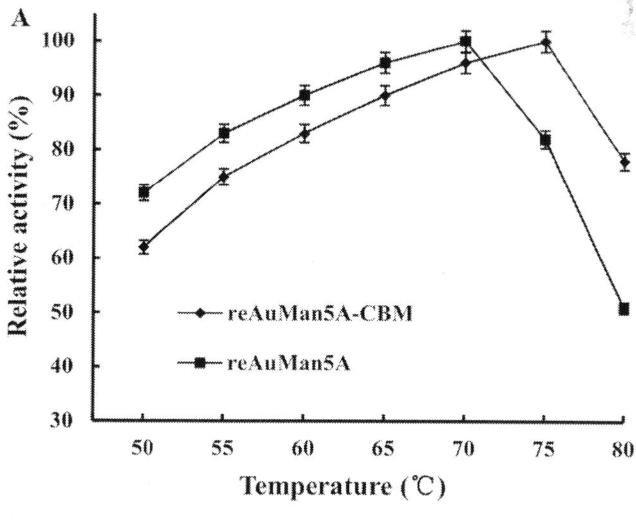

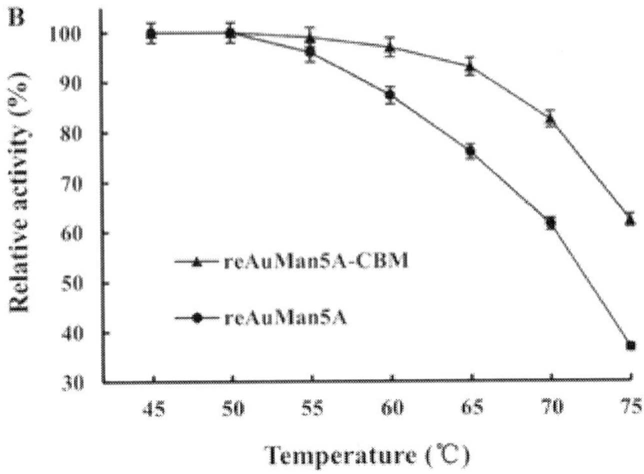


**Figure 5:** Effects of temperature on (A) activity and (B) stability of the reAuMan5A-CBM and reAuMan5A.

Temperature optima of the reAuMan5A-CBM and reAuMan5A were determined, respectively, under the standard assay conditions, except temperatures ranging from 50 to 80°C. Their thermostabilities were evaluated, respectively, by incubating them at pH 3.6 and various temperatures from 45 to 75°C for 1.0 h, and the residual enzyme activities were measured under the standard assay conditions.

https://doi.org/10.1371/journal.pone.0064766.g005

The $K_m$ and $V_{max}$ values of the reAuMan5A-CBM were determined to be 0.66 mg/mL and 389.1 U/mg, respectively, and those of the reAuMan5A were 1.36 mg/mL and 415.8 U/mg, respectively. The reAuMan5A-CBM displayed an obvious decrease in $K_m$ but a slight alteration in $V_{max}$, as compared with the reAuMan5A.

## Cellulose-binding Capacity

A total of 20 mL of dialyzed cultured supernatant, in which the reAuMan5A-CBM or reAuMan5A activity was adjusted to 20 U/mL, was loaded onto the cellulose column. The test results showed that cellulose-binding capacity of the reAuMan5A-CBM with Avicel PH-101 was up to 92.3%, whereas that of the reAuMan5A was not detected.

# DISCUSSION

Mannans, the major hemicelluloses in plant cell walls and the specific substrates for β-mannanases, are always cross-linked with various kinds of carbohydrates, such as celluloses, xylans and arabinans [19]. CBMs often existed in some hydrolases decomposing plant cell walls, such as β-mannanases, xylanases and arabino furanosidases. Being natively in enzymes or by fusing CBMs into the N- or C-termini of other enzymes, the CBMs could obviously enhance the catalytic efficiency of enzymes by increasing their local concentrations around the polysaccharide substrates, and/or improve the thermostability [17]. Because the AuMan5A consists of an only CD, it is reasonable to perfect its enzymatic properties, such as thermostability and cellulose-binding capacity, by fusing a CBM into the AuMan5A. However, there are a total of 25,580 CBMs classified into 64 families in the CAZY database, so the choice of a suitable CBM from so numerous CBMs is time-consuming and laborious. It was reported that the family 1 CBMs are mainly cellulose-binding modules, while celluloses are often covalently and non-covalently attached to mannans [5]. Therefore, our present work was focused on the family 1 CBMs.

The MD simulation can rapidly predict the most suitable binding position for a ligand on a given macromolecule and calculate the binding free energy, which was used as a powerful tool in indirectly reflecting the affinity between ligand and receptor, such as the design of human phospholipase $A_2$ or HIV-1 protease inhibitors [20], [21]. In this work, we selected 23 family 1 CBMs to construct the phylogenetic tree, from which 8 candidate CBMs were chosen. And their 3-D structures were modeled based on a CBM crystal one of the TrCBH I. Next, the interaction between 3-D structures of the candidate CBM and cellobiose was predicted, and the binding free energy was calculated. And finally, a CBM having the lowest binding free energy was selected, and then the *Auman5A-cbm* was constructed and expressed in *P. pastoris* GS115. As can be seen from Fig. 2, several aromatic amino acids are exposed to the groove surface of a CBM of the TrCBH I. In recent years, the 3-D structures of representative members from 22 CBM families have been resolved. Data from these 3-D structures speculated that these aromatic amino acids may play a key role in binding cellobiose or carbohydrates [22], which needs to be proved by our further studies.

*P. pastoris* transformant that can resist a higher concentration of G418 might have multiple copies of integration of a heterologous gene into *P. pastoris* genome, which could potentially result in a high expression level of a heterologous protein as explained in the manual of Multi-Copy Pichia Expression Kit (Invitrogen). However, the expression level was not directly proportional to the concentration of G418 [9], [23]. Due to those reasons, in this work, a total of 30 transformants with *Auman5A-cbm* resistant to different concentrations of G418 and 30 transformants with *Auman5A* were picked out for flask expression tests. This screening procedure has been applied to conduct the over-expression of other recombinant proteins or enzymes in *P. pastoris* [24], [25].

The *P. pastoris* expression system has many advantages, one of which is that the purity of expressed recombinant proteins or enzymes is high. It was reported that the purity of recombinant *A. sulphureus* β-mannanase expressed in *P. pastoris* X-33 was 97% [9]. In this work, the purity of reAuMan5A-CBM or reAuMan5A was more than 85%. So they were purified to homogeneity only by a simple combination of ammonium sulfate precipitation, ultrafiltration and Sephadex G-75 gel filtration (Fig. 4). The purified reAuMan5A-CBM and reAuMan5A were N-glycosylated proteins, which were confirmed by N-deglycosylation and carbohydrate content assays.

The temperature optimum and stability of the reAuMan5A-CBM were 75 and 68°C, which were 5 and 8°C higher than those of the reAuMan5A, respectively, indicating that the fusion of a CBM of the TrCBH I conferred the elevated tolerance to high temperature on the AuMan5A. The $K_m$ value (0.66 mg/mL) of the reAuMan5A-CBM was much lower than that (1.36 mg/mL) of the reAuMan5A, implying that the former has a higher affinity towards locust bean gum. The cellulose-binding capacity of the reAuMan5A-CBM with Avicel PH-101 was up to 92.3%, suggesting that it can be gathered around the natural substrate (that is, mannan cross-linked with cellulose, xylan and arabinan) by exclusively binding cellulose [26].

## CONCLUSIONS

In this work, the fusion β-mannanase gene, *Auman5A-cbm*, was constructed as it was designed theoretically, and functionally expressed in *P. pastoris* GS115. The fusion of a family 1 CBM of the TrCBH I into the C-terminus of AuMan5A obviously improved its thermostability and cellulose-binding capacity, which are beneficial for industrial applications. The superior properties of the reAuMan5A-CBM make it a good candidate in industrial processes. To our knowledge, this work first established a novel strategy for molecular modification of enzymes by *in silico* design.

## ACKNOWLEDGMENTS

We are grateful to Prof. Xianzhang Wu (School of Biotechnology, Jiangnan University) for providing technical assistance.

## AUTHOR CONTRIBUTIONS

Conceived and designed the experiments: MCW JFL CDT. Performed the experiments: CDT XHW. Analyzed the data: RM SJG JQW. Contributed reagents/materials/analysis tools: XY. Wrote the paper: CDT MCW.

## REFERENCES

1. van Zyl WH, Rose SH, Trollope K, Görgens JF (2010) Fungal β-mannanases: Mannan hydrolysis, heterologous production and biotechnological applications. Proc Biochem 45: 1203–1213.
2. Heck JX, de Barros Soares LH, Ayub MAZ (2005) Optimization of xylanase and mannanase production by *Bacillus circulans* strain BL53 on solid-state cultivation. Enzyme Microb Technol 37: 417–423.
3. Wu MC, Tang CD, Li JF, Zhang HM, Guo J (2011) Bimutation breeding of *Aspergillus niger* strain for enhancing β-mannanase production by solid-state fermentation. Carbohydr Res 346: 2149–2155.
4. Roth R, Moodley V, van Zyl P (2009) Heterologous expression and optimized production of an *Aspergillus aculeatus* endo-1,4-β-mannanase in *Yarrowia lipolytica*. Mol Biotechnol 43: 112–120.
5. Guillén D, Sánchez S, Rodríguez-Sanoja R (2010) Carbohydrate-binding domains: multiplicity of biological roles. Appl Microbiol Biotechnol 85: 1241–1249.

6. Ravalason H, Herpoël-Gimbert I, Record E, Bertaud F, Grisel S, et al. (2009) Fusion of a family 1 carbohydrate binding module of *Aspergillus niger* to the *Pycnoporus cinnabarinus* laccase for efficient softwood kraft pulp biobleaching. J Biotechnol 142: 220–226.
7. Zhang Y, Chen S, Xu M, Cavaco-Paulo A, Wu J, et al. (2010) Characterization of *Thermobifida fusca* cutinase-carbohydrate-binding module fusion proteins and their potential application in bioscouring. Appl Environ Microbiol 76: 6870–6876.
8. Zhang YL, Ju JS, Peng H, Gao F, Zhou C, et al. (2008) Biochemical and structural characterization of the intracellular mannanase AaManA of *Alicyclobacillus acidocaldarius* reveals a novel glycoside hydrolase family belonging to clan GH-A. J Biol Chem 283: 31551–31558.
9. Chen XL, Cao YH, Ding YH, Lu WQ, Li DF (2007) Cloning, functional expression and characterization of *Aspergillus sulphureus* β-mannanase in *Pichia pastoris*. J Biotechnol 128: 452–461.
10. Stalbrand H, Saloheimo A, Vehmaanpera J, Henrissat B, Penttila M (1995) Cloning and expression in *Saccharomyces cerevisiae* of a *Trichoderma reesei* β-mannanase gene containing a cellulose binding domain. Appl Environ Microbiol 61: 1090–1097.
11. Benech RO, Li XM, Patton D, Powlowski J, Storms R, et al. (2007) Recombinant expression, characterization, and pulp prebleaching property of a *Phanerochaete chrysosporium* endo-β-1,4-mannanase. Enzyme Microb Technol 41: 740–747.
12. Tang CD, Guo J, Wu MC, Zhao SG, Shi HL, et al. (2011) Cloning and bioinformatics analysis of a novel acidophilic β-mannanase gene, *Auman5A*, from *Aspergillus usamii*YL-01–78. World J Microbiol Biotechnol 27: 2921–2929.
13. Timmers LF, Ducati RG, Sanchez-Quitian ZA, Basso LA, Santos DS, et al. (2012) Combining molecular dynamics and docking simulations of the cytidine deaminase from *Mycobacterium tuberculosis* H37Rv. J Mol Model 18: 467–479.
14. Phosrithong N, Ungwitayatorn J (2010) Molecular docking study on anticancer activity of plant-derived natural products. Med Chem Res 19: 817–835.

15. Li JF, Zhao SG, Tang CD, Wang JQ, Wu MC (2012) Cloning and functional expression of an acidophilic β-mannanase gene (*Anman5A*) from *Aspergillus niger* LW-1 in *Pichia pastoris*. J Agric Food Chem 60: 765–773.
16. Laemmli UK (1970) Cleavage of structural proteins during the assembly of the head of bacteriophage T4. Nature 227: 680–685.
17. Thongekkaew J, Ikeda H, Iefuji H (2012) Increases thermal stability and cellulose-binding capacity of *Cryptococcus* sp. S-2 lipase by fusion of cellulose binding domain derived from *Trichoderma reesei*. Biochem Biophys Res Commun 420: 183–187.
18. Dubois M, Gilles KA, Hamilton JK, Rebers PA, Smith F (1956) Colorimetric method for determination of sugars and related substances. Anal Chem 28: 350–356.
19. Moreira LRS, Filho EXF (2008) An overview of mannan structure and mannan-degrading enzyme systems. Appl Microbiol Biotechnol 79: 165–178.
20. Ortiz AR, Pisabarro MT, Gago F, Wade RC (1995) Prediction of drug binding affinities by comparative binding energy analysis. J Med Chem 38: 2681–2691.
21. Perez C, Pastor M, Ortiz AR, Gago F (1998) Comparative binding energy analysis of HIV-1 protease inhibitors: incorporation of solvent effects and validation as a powerful tool in receptor-based drug design. J Med Chem 41: 836–852.
22. Shoseyov O, Shani Z, Levy I (2006) Carbohydrate binding modules: biochemical properties and novel applications. Microbiol Mol Biol Rev 70: 283–295.
23. Li JF, Tang CD, Shi HL, Wu MC (2011) Cloning and optimized expression of a neutral endoglucanase gene (*ncel5A*) from *Volvariella volvacea* WX32 in *Pichia pastoris*. J Biosci Bioeng 111: 537–540.
24. Tan ZB, Li JF, Wu MC, Tang CD, Zhang HM, et al. (2011) High-level heterologous expression of an alkaline lipase gene from *Penicillium cyclopium* PG37 in *Pichia pastoris*. World J Microbiol Biotechnol 27: 2767–2774.

25. Shi HL, Yin X, Wu MC, Tang CD, Zhang HM, et al. (2012) Cloning and bioinformatics analysis of an endoglucanase gene (*Aucel12A*) from *Aspergillus usamii* and its functional expression in *Pichia pastoris*. J Ind Microbiol Biotechnol 39: 347–357.
26. Ye XH, Zhu ZG, Zhang CM, Zhang YHP (2011) Fusion of a family 9 cellulose-binding module improves catalytic potential of *Clostridium thermocellum* cellodextrin phosphorylase on insoluble cellulose. Appl Microbiol Biotechnol 92: 551–560.

CHAPTER 8

# COMBINED ENZYMATIC AND PHYSICAL DEINKING METHODOLOGY FOR EFFICIENT ECO-FRIENDLY RECYCLING OF OLD NEWSPRINT

**Antar Puneet Virk[1] , Minakshi Puri[2] , Vijaya Gupta[2] , Neena Capalash[1] , Prince Sharma[2]**

[1]Department of Biotechnology, Panjab University, Chandigarh, India

[2]Department of Microbiology, Panjab University, Chandigarh, India

## ABSTRACT

### Background

The development in the deinking process has made recycled fiber a major part of the raw material for pulp and paper industry. Enzymes have revolutionized the deinking process obtaining brightness levels surpassing conventional deinking processes. This study explores the deinking efficiencies of bacterial alkalophilic laccase (L) and xylanase (X) enzymes along with physical deinking methods of microwaving (MW) and sonication (S) for recycling of old newsprint (ONP).

## Methods and Results

The operational parameters viz. enzyme dose, pH and treatment time for X and L deinking were optimized statistically using Response Surface Methodology. Laccase did not require any mediator supplementation for deinking. Deinking of ONP pulp with a combination of xylanase and laccase enzymes was investigated, and fiber surface composition and morphological changes were studied using X-ray diffraction, fourier transform infrared spectroscopy and scanning electron microscopy. Compared to the pulp deinked with xylanase (47.9%) or laccase (62.2%) individually, the percentage reduction of effective residual ink concentration (ERIC) was higher for the combined xylanase/laccase-deinked pulp (65.8%). An increase in brightness (21.6%), breaking length (16.5%), burst factor (4.2%) tear factor (6.9%), viscosity (13%) and cellulose crystallinity (10.3%) along with decrease in kappa number (22%) and chemical consumption (50%) were also observed. Surface appeared more fibrillar along with changes in surface functional groups. A combination of physical and enzymatic processes (S-MW-XL) for deinking further improved brightness (28.8%) and decreased ERIC (73.9%) substantially.

## Conclusion

This is the first report on deinking of ONP with laccase without any mediator supplementation. XL pretreatment resulted in marked improvement in paper quality and a new sequence being reported for deinking (S-MW-XL) will contribute further in decreasing chemical consumption and making the process commercially feasible.

# INTRODUCTION

Recycling of wastepaper has gained momentum over the past decades due to the severity in the demand of green plants being imposed by the paper industry throughout the world [1]. Deinking is an important step in the recycling process and involves the dislodgement of ink particles from fiber surface and then removal of the detached ink particles by flotation, washing etc [2]. The developments in the deinking process have immensely helped the utilization of secondary fiber such as old newsprint, xeroxed papers and laser/inkjet printed papers for making white grade papers. Out of the two types of inks present in secondary fibers, it is easier to remove impact ink such as the one in old newsprint as compared to the non-impact ink present in xeroxed or laser/inkjet printed papers.

The conventional methods of deinking utilize chemicals such as sodium hydroxide, sodium silicate, hydrogen peroxide, chlorine based chemicals and other chelating agents which are environmentally hazardous [3]. Enzymatic deinking has attracted enormous attention by reducing the consumption of chemicals thereby lowering the process cost and making it eco-friendly [2]. Enzymes can act either directly on the fiber or on the ink. Different enzymes being employed for deinking include cellulases, xylanases, pectinases, amylases, lipases, esterases and laccases [4]. The enzymatic deinking using cellulases and hemicellulases alone and in combination has been well characterized [5]. Owing to their action on the cellulosic content of the paper lowering the strength properties, it is desirable to search for more effective enzymes. Laccases are oxidative enzymes that influence the pulp and paper industry in several ways [6]. The role of laccases has been well established in the delignification of pulps resulting in an increase in viscosity [7], [8]. Hence, laccases could be proposed as interesting alternative for deinking of old newsprint which chiefly contains lignin rich mechanical pulp. The detachment of ink particles as well as contaminants, with additional chemical modifications induced by laccase enzyme, could enhance brightness and reduce residual ink concentration [9]. There are reports on use of laccase-mediator systems alone and in combination with cellulases and hemicellulases [10], [2]. The laccases used till date for deinking purposes have been obtained from either fungal sources [5] or plants [9].

Physical methods such as sonication and high temperature are pretreatment stages for deinking owing to their ability to reduce chemical consumption and enhance physical as well as optical properties of recycled papers [11]. As compared to the conventional physical methods using conduction/convection heating, microwaving allows direct interaction between a heated object and an applied electromagnetic field to create volumetric and rapid heating which causes an explosion effect among particles, and improves disruption of recalcitrant structures [12]. Microwave pretreatment was used previously for the biobleaching of kraft pulp in our laboratory [13], and was used for deinking purposes in this work.

This study employed bacterial laccase and xylanase enzymes, sonication and microwaving, alone and in combination, for the deinking of ONP pulp. The parameters for enzymatic deinking were optimized statistically. The optical and physico-mechanical properties of deinked pulp and reduction in chemical consumption were evaluated. The surface properties of enzymatically treated pulp were studied by X-ray diffraction, fourier transform infrared spectroscopy and scanning electron microscopy.

# MATERIALS AND METHODS

## Wastepaper Pulps

The different pulps used were made from locally procured wastepapers viz. old newsprint, magazines, laser, inkjet and xerox with initial brightness of 36, 42, 65, 60 and 58% ISO respectively.

## Enzymes

A cellulase-free, thermo-alkali-stable and halo-tolerant xylanase with optimum activity at pH 8–9.5 and 65°C was produced from *Bacillus halodurans* FNP 135 soil isolate by solid state fermentation of wheat bran under optimized conditions viz. 2.31% $Na_2CO_3$ (w/w), 80% moisture content (v/w), 68 h of fermentation at 37°C and assayed using birchwood xylan as substrate [14]. One unit activity was defined as the amount of enzyme that produced 1 μmol of reducing sugar (expressed as xylose) per min.

A cellulase and xylanase-free laccase with optimum activity at pH 8 and 55°C was produced from *gamma-proteobacterium,* a novel bacterium isolated from industrial effluent [15] and later identified as a new *Rheinheimera* species [16]. One ml of overnight culture was used to inoculate 100 ml of M162 medium [17] containing 100 μM copper and incubated at 37°C at 150 rpm for 48 h. The culture was centrifuged at 10,000×*g*, 4°C for 10 min and the supernatant was used as crude enzyme preparation. Laccase activity was determined using 2 mM guaiacol, at

55°C in 0.1 M phosphate buffer (pH 6.5) [15]. The change in absorbance due to oxidation of substrate in the reaction mixture was monitored after 10 min of incubation at 465 nm ($\varepsilon$=48000 $M^{-1}cm^{-1}$). Unit of laccase was expressed in nkat (n moles of substrate converted/sec/ml of enzyme).

## Pulp Preparation

Waste papers were shredded and soaked overnight in tap water at room temperature. The soaked papers were washed several times and were disintegrated with the help of a grinder to obtain soft cottony pulp. It was then squeezed to remove absorbed water and oven dried. This oven dried pulp (odp) was used for further experiments.

## Pulp Deinking Sequences

All treatments were carried out in triplicates at least. A standard deinking sequence (P-C) was performed in the laboratory. P refers to the pulp preparation stage and C refers to chemical bleaching stage where pulp at 5% consistency was treated with sodium hydroxide (2%), sodium silicate (2.5%), DTPA (Diethylene triamine penta-acetate, 0.5%), Triton X-100 (1.2%) and $H_2O_2$(1%) at 50°C and 200 rpm for 15 min. To facilitate removal of ink particles detached from fibers, pulp was thoroughly washed with distilled water and was used to make hand sheets to determine brightness of the pulp.

Enzymatic and physical pretreatments were evaluated using the standard sequence (P-C) by incorporating (a) xylanase (X) stage (P-X-C), (b) laccase (L) stage (P-L-C), (c) L stage along with mediators/inhibitors (M/I) (P-L+M/I-C), (d) X followed by L stage (P-X-L-C), (e) sonication followed by XL stage (P-S-X-L-C), (f) microwaving (MW) followed by XL stage (P-MW-X-L-C), and (g) S followed by MW followed by XL stage (P-S-MW-X-L-C). In all cases, standard sequence (P-C) was used as a control. Both xylanase and laccase treatments were carried out with 10 g (dry weight) of pulp at 5% consistency at different enzyme doses, temperatures and pH for optimization of various parameters for deinking using Response Surface Methodology. The treatments were carried out in flasks, at 160 rev/min and 65°C and 55°C for xylanase and laccase respectively. Laccase mediator and inhibitor treatments were carried out with 10 g (dry weight) of pulp at 5% consistency at optimum pH, enzyme dose and incubation time using 2 mM of 2,2′-azino-bis-3-ethylbenzothiazoline-6-sulphonic acid (ABTS), 1-hydroxybenzotriazole (HBT), syringaldehyde (Syr) as mediators and 1 mM of sodium azide (SA) as inhibitor. Sonication [11] was carried out at 1% consistency, 20 kHz and microwaving [13] was carried out at 20% consistency, 850 W employing optimum pH, enzyme dose and incubation time for different time intervals.

## Analysis of Collected Filtrate

The color removal from the pulp was determined spectrophotometrically from λ 200 nm to λ 800 nm. The reducing sugar released was measured by the dinitrosalicylic acid (DNS) [18], phenolic compounds generated were estimated by measuring the absorbance at λ 237 nm and release of hydrophobic compounds were determined by measuring the absorbance at λ 465 nm [19], [20].

## X-ray Diffraction

Crystallinity of cellulose was determined for both untreated and XL treated samples by X-ray diffraction. After freeze drying, samples were analyzed using the X-ray diffractometer with X-ray generator from 2 to 50 of 2θ (scattering angle). The crystalline index of cellulose ($X_c$) was calculated from the X-ray diffraction patterns by the following equation [21].

$$X_C = \frac{I_{002} - I_{am} \times 100\%}{I_{002}}$$

Where $I_{0\,0\,2}$ is the peak intensity from the (0 0 2) lattice plane (2θ=22.6°) and $I_{am}$ is the peak intensity of amorphous phase (2θ=19.0°). Apparent crystallite size (ACS) was estimated through the use of the Scherrer equation [21].

$$ACS = 0.89 \times \lambda\beta\cos\theta$$

where λ is the wavelength of the incident X-ray (1.5418 Å), θ the Bragg angle corresponding to the (002) plane, and β the half-height width of the peak angle of the (002) reflection.

## Fourier Transformed Infrared Spectroscopy (FTIR)

FTIR spectra for pulp samples were recorded with a resolution of 4 $cm^{-1}$ over the wave number range of 4000–400 $cm^{-1}$, using 32 scans per sample. Empirical cellulose crystallinity index (CI) was calculated by dividing absorbance intensities of IR bands at 1430 and 897 $cm^{-1}$ [22].

$$CI = \frac{A_{1430}}{A_{897}}$$

FTIR spectrum was analyzed using the FTIR database [23].

## Scanning Electron Microscopy of Pulp Fibers

The fibers were washed thrice with deionized water and were gradually dehydrated with acetone gradient between 30 and 90% and finally suspended in 100% acetone; small pieces of fibers were air dried and placed on the stubs, mounted with silver tape, and sputter coated with gold using fine coat (JEOL ion sputter, Model JFC-1100) and examined at 10 KV.

## Physical and Chemical Characterization of Hand Sheets

All physical, chemical and residual chlorine measurements were carried out by following the TAPPI standard methods for brightness and whiteness (T-452 OM-87 and T-1216 OM respectively with a Minolta spectrophotometer CM3630), kappa number (T 236), tear factor (T-414 OM-98), burst factor (T-403 OM-97), breaking length (T-404 OM-92) and viscosity (T-230 OM-82). The effective residual ink concentration was measured with the Elrepho ERIC tester using Lorentzen and Wettre software.

# RESULTS AND DISCUSSION

## Optimization of Deinking of ONP Pulp with Xylanase (P-X-C) and Laccase (P-L-C) Using Response Surface Methodology

Central composite design with three factors - pH, enzyme dose and incubation time at five levels was employed to investigate the first and higher order main effects of each factor and interaction amongst them for deinking of ONP pulp using xylanase and laccase enzymes. With xylanase, maximum brightness of 53.5±0.4% ISO (9% increase) at temperature 65°C, pH 9.0, enzyme dose 15 U/g odp, 150 rpm and incubation time 3 h was obtained; With laccase, maximum brightness of 54.9±0.5% ISO (12% increase) at temperature 55°C, pH 8.0, enzyme dose 50 U/g odp, 150 rpm and incubation time 4 h was obtained. This is in accordance with previous studies which showed a brightness increase of 3.22 and 1.7% [10]and 19.4 and 7.8% [5] after xylanase and laccase-mediator pretreatments respectively. ANOVA results of the data disclosed that the models, all independent variables, interactions among pH, enzyme dose as well as time were significant and the lack of fit was not significant. The 3D graphs showed that brightness of deinked pulp was affected by the variation of two factors at a time keeping the third at optimal condition in both xylanase and laccase pretreatments and maximum brightness was obtained when all three factors were at their optimal levels.

## Mediator Supplementation for Deinking with Laccase

Different laccase mediators were tried to further enhance deinking (Figure 1). Except for ABTS, no other mediator showed any deinking effect either alone or in combination with laccase. However, ABTS supplementation showed only an additive effect when used with laccase rather than showing a mediating effect. Therefore, the bacterial laccase used in this study is being reported to require neither ABTS nor any other mediator for deinking, contrary to the previous reports [5], [11] where ABTS was found essential as a mediator for effective deinking of ONP pulp with laccase. ABTS acts as an electron shuttle between pulp fiber and the laccase molecule [24] and thereby causes deinking by oxidative action. A laccase requires a mediator only when its redox potential is less than that of the substrate to be oxidized. If the redox potential of laccase is already higher than the substrate (ink), then the need for mediator is eliminated.

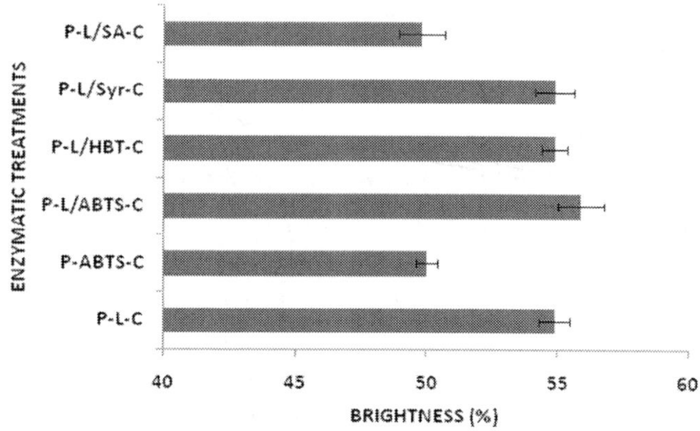


**Figure 1:** Effect of mediators (ABTS - 2,2′-azino-bis-3-ethylbenzothiazoline-6-sulphonic acid, HBT - 1-hydroxybenzotriazole, Syr - syringaldehyde) and inhibitor (SA – sodium azide) on the deinking of ONP pulp with laccase enzyme.

https://doi.org/10.1371/journal.pone.0072346.g001

Chemical deinking causes removal (dislodging) of ink particles from pulp fiber only whereas laccase enzyme may have caused either decolourization only or both decolourization and dislodging of ink particles. This was further supported by the fact that on comparing the absorbance scans of effluents (λ 200–800 nm) to detect the release of colored compounds, chemically treated pulp effluent showed high absorbance whereas laccase treated and untreated (control) pulp effluents were colourless.

This is the first report on deinking of ONP pulp with a bacterial laccase without the need of a mediator, thus making the process cost effective and eliminating mediator-linked enzyme toxicity problems. Furthermore, no deinking effect was observed with laccase in the presence of laccase inhibitor sodium azide which confirmed the role of laccase in deinking (Figure 1).

## Sequential Deinking of ONP Pulp with Xylanase and Laccase (P-X-L-C)

The present work demonstrated that an improved level of deinking was achieved by using xylanase and laccase sequentially as compared to xylanase or laccase alone. Pretreatment of ONP pulp with xylanase, followed by laccase, obtained a brightness of 59.6±0.8% ISO which was

8.5, 11.4 and 21.6% higher than that of the P-L-C, P-X-C and untreated pulp (Table 1). ERIC for P-X-L-C deinked pulp was 183.07±24 ppm which was 9.4, 35.2 and 65.8% lower than P-L-C, P-X-C and untreated pulp. The substantial reduction in residual ink concentration clearly showed the effective role of laccase followed by xylanase in deinking of ONP pulp. Xylanase acts on waste pulp, thereby removing xylan and opening the fibers for the subsequent chemical or enzyme attack. Laccase acts on the exposed lignin content of the pulp thereby facilitating the removal of lignin monomers and contaminants attached to lignin besides removing/dislodging ink particles. This synergistic action of xylanase and laccase thereby enhanced the brightness. Bacterial laccases have been previously reported to show potential in the biobleaching of pulp [25], [7], but not for their use in deinking of pulp. The correlation between release of chromophores (λ 237 nm), hydrophobic compounds (λ 465 nm), and reduction in kappa number coupled to the release of reducing sugars suggested the dissociation of lignin and sugars from pulp fibers (Figure 2). This work reports for the first time the use of a bacterial laccase alone and in combination with bacterial xylanase for deinking of pulp. The combined enzymatic treatment caused a substantial (50%) reduction in chemical consumption for deinking of ONP pulp (Table 2).

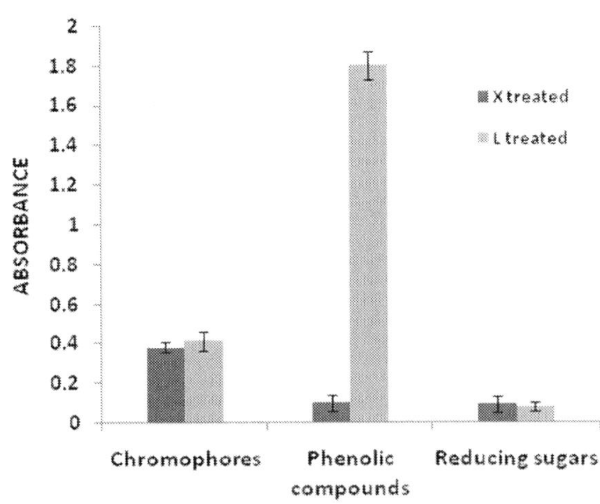


**Figure 2:** Analysis of phenolic compounds, hydrophobic compounds and reducing sugars in effluent released after xylanase (X) and laccase (L) treatment.

https://doi.org/10.1371/journal.pone.0072346.g002

**Table 1**: **Physical and Chemical characterization of enzymatica**lly and physically deinked pulp

https://doi.org/10.1371/journal.pone.0072346.t001

| Parameters | PC | PXC | PLC | PXLC | PSXLC | PMWXLC | PSMWXLC |
|---|---|---|---|---|---|---|---|
| Kappa number | 10±0.9 | 9±0.78 | 8.8±0.74 | 8.1±0.83 | 7.8±0.75 | 7.7±0.75 | 7.4±0.66 |
| Brightness (% ISO) | 49.0±0.5 | 53.5±0.4 | 54.9±0.5 | 59.6±0.8 | 61.45±0.4 | 61.73±0.4 | 63.12±0.5 |
| Whiteness | 1.31±0.6 | 9.75±0.8 | 10.3±0.8 | 22.7±0.5 | 23.85±0.6 | 24.09±0.9 | 25.47±0.8 |
| ERIC (ppm) | 535.41±16 | 278.55±23 | 202.24±33 | 183.07±24 | 145.20±30 | 148.51±16 | 139.46±22 |
| Breaking Length (m) | 1285±29 | 1182±25 | 1378±22 | 1498±25 | 1328±32 | 1331±30 | 1345±33 |
| Burst Factor (kPa $m^2$/g) | 6.12±0.8 | 6.34±0.56 | 6.27±0.62 | 6.38±0.58 | 6.24±0.9 | 6.26±0.7 | 6.3±0.5 |
| Tear Factor (mN $m^2$/g) | 37.2±0.20 | 35.5±0.23 | 38.5±0.20 | 39.8±0.26 | 34.5±0.25 | 35.1±0.20 | 37.8±0.20 |
| Viscosity (cps) | 7.5±0.8 | 7.8±0.9 | 8.0±0.6 | 8.5±0.9 | 7.9±0.7 | 8.0±0.8 | 8.65±0.9 |

doi:10.1371/journal.pone.0072346.t001

**Table 2:** Deinking of waste paper pulps by P-X-L-C treatment and reduction in chemical load

https://doi.org/10.1371/journal.pone.0072346.t002

| Waste paper pulp | | Brightness (%ISO) | | | | | | |
|---|---|---|---|---|---|---|---|---|
| | P-C | P-X-L-C | | | | | | |
| Chemicals used | 100% | 100% | 90% | 80% | 70% | 60% | 50% | 40% |
| ONP print | 49±0.5 | 59.6±0.8 | 58.9±0.9 | 56.5±0.5 | 54.7±0.8 | 52.5±0.73 | 49.2±0.65 | 48.3±0.75 |
| Laser print | 81.5±0.4 | 84.1±0.5 | 82.8±0.8 | 82.3±0.3 | 81.5±0.7 | 79.2±0.6 | - | - |
| Magazines print | 51±0.5 | 55.24±0.7 | 52.5±0.9 | 52.2±0.5 | 50.9±0.5 | 48.9±0.4 | - | - |
| Xerox print | 85.6±0.4 | 87.3±0.5 | 86.8±0.4 | 86.1±0.8 | 85.7±0.7 | 84.1±0.4 | - | - |
| Inkjet print | 82±0.9 | 85.84±0.5 | 83.8±0.6 | 82.2±0.5 | 80.8±0.4 | - | - | - |

doi:10.1371/journal.pone.0072346.t002

# X-ray Diffraction Analysis

The X-ray diffraction analysis of xylanase–laccase (P-X-L-C) treated ONP pulp is shown in Figure 3. The peak intensity from (002) lattice plane (2θ=22.6°) represents the crystalline cellulose, while peak intensity of amorphous phase occur at 2θ=19°. The $I_{002}$ intensities (in cm) at 2θ=22.6° were 2.4 and 8.4, $I_{am}$ intensities (in cm) at 2θ=19° were 1.7 and 2.9, half height width (in cm) of the peak angle of the (002) reflection at 2θ=22.6° was 1.9 and 1.1 for the P-C and P-X-L-C pulps respectively. XRD analysis revealed that P-X-L-C treatment increased the crystallinity index of the ONP pulp from 29.16% (untreated) to 65.47% while the cellulose crystallinity index increased by 10.3% (Table 3) which means a decrease in amorphous cellulose and increase in crystalline cellulose of the pulp [26]. The increase in crystallinity might be due to removal of hemicelluloses and lignin and components adhered to lignin as a result of P-X-L-C treatment thereby increasing the cellulose content of the pulp [27]. The apparent cellulose crystal size of P-X-L-C treated pulp was altered by enzyme treatment showing 73% increase.

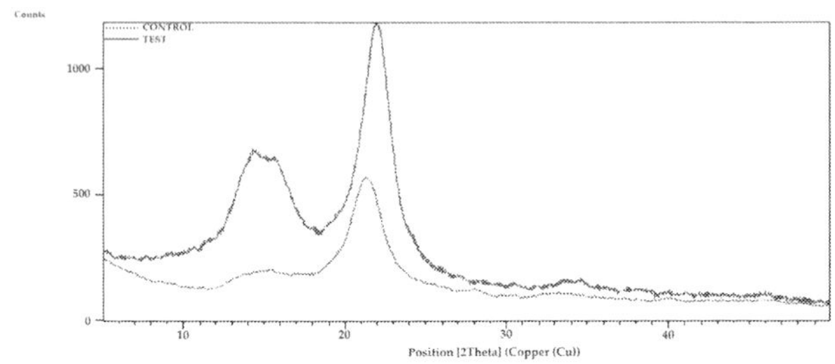


**Figure 3:** X-ray diffraction analysis of control (P-U-C) and test (P-X-L-C treated) ONP pulp.

https://doi.org/10.1371/journal.pone.0072346.g003

**Table 3**: **The crystallinity index and apparent crystal size (ACS) of untreated (P-C) and treat**ed (P-X-L-C) ONP pulp cellulose

https://doi.org/10.1371/journal.pone.0072346.t003

| Parameters | P-C | P-X-L-C |
|---|---|---|
| Empirical crystalline index | 29.16 | 65.47 |
| Apparent crystal size (nm) | 0.78 | 1.35 |
| Cellulose crystallinity index (CI) | 1.16 | 1.28 |

doi:10.1371/journal.pone.0072346.t003

## FTIR Spectral Analysis

FTIR spectra for xylanase–laccase (P-X-L-C) treated pulp samples showed several characteristic and prominent changes (Figure 4). The P-X-L-C treated pulp showed absorption structures similar to P-C but with different intensities (Table 4). The bands around 3401–3352 $cm^{-1}$ depict –OH stretching of hydrogen-bonding. The increase in their relative intensity after enzymatic treatments is attributed to the increase in cellulosic content of the pulp. The decrease in relative intensity of bands at 2921–2917 $cm^{-1}$ and 617 $cm^{-1}$, assigned to CH asymmetrical stretching vibration in $CH_3$, $CH_2$, CH, in P-X-L-C treated pulp indicated the degradation of aliphatic side chains. Relative intensity around 1642–1639 $cm^{-1}$ increased in P-X-L-C treated pulp which is attributed to release of free carbonyl groups (C=O) due to action of enzyme on lignin's aromatic ring [23]. In P-C pulp, these carbonyl groups remain associated with aromatic rings resulting in less absorption. Carboxylic acids appear to be intermediate degradation products in the complete mineralization of lignins to carbon dioxide and water [28]. The relative intensity of P-X-L-C pulp decreased at 1456–1426 $cm^{-1}$, a band assigned to aromatic skeletal vibrations combined with $–CH_3$ in-plane deformations, showing that some methoxyl groups were removed during the enzymatic treatment. The band at 1374–1372 $cm^{-1}$was assigned to aliphatic C–H stretching in $CH_3$ (not in $–OCH_3$) and phen–OH. The decrease of its relative intensity illustrated that either the side chains or phen–OH of lignin decreased after the enzymatic treatment. The decrease in relative intensity at 1266–1252 $cm^{-1}$ indicates degradation of guaiacyl groups. A new band that appeared at 1737 $cm^{-1}$ in P-X-L-C treated pulp was assigned to C=O stretching vibration in β-C=O, COOH, ester indicating that residual lignin after an P-X-L-C treatment was enriched in these types of functional groups. New band at 1162 $cm^{-1}$ in P-X-L-C treated pulp indicated degradation of syringyl groups. The changes in pulp crystallinity derived from $A_{1430}/A_{897}$ ratio in accordance with the X-ray diffraction studies.

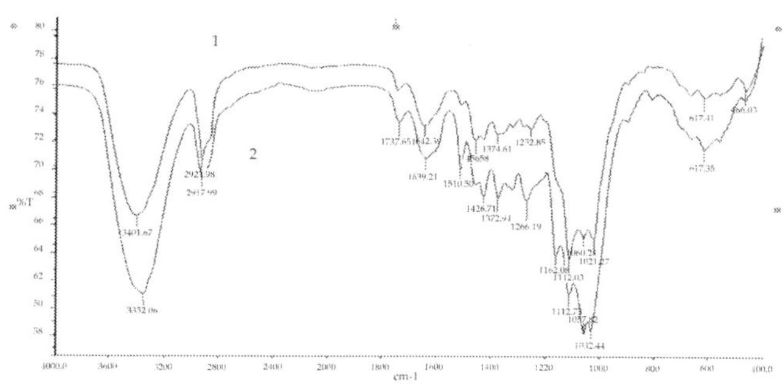


**Figure 4:** FTIR spectra of (1) P-C (2) P-X-L-C treated ONP pulp.

https://doi.org/10.1371/journal.pone.0072346.g004

**Table 4:** Signals assignment of hand sheets in FTIR spectra and their relative intensity for untreated (P-C) and treated (P-X-L-C) pulp

https://doi.org/10.1371/journal.pone.0072346.t004

| Bands $cm^{-1}$ | Assignment | Relative intensity | |
|---|---|---|---|
| | | P-C | P-X-L-C |
| 3401–3352 | –OH stretching of hydrogen-bonding | 1.28 | 1.36 |
| 2921–2917 | CH asymmetrical stretching vibration in $CH_3$, $CH_2$, CH | 1.06 | 0.95 |
| 1737* | C=O stretching vibration in β-C=O, COOH, ester | - | 0.94 |
| 1642–1639 | Aromatic skeletal vibrations plus C=O stretching | 1.07 | 0.98 |
| 1510 | Aromatic skeletal vibrations | 1 | 1 |
| 1456–1426 | Aromatic skeletal vibrations combined with $-OCH_3$ in plane deformations | 1.111 | 1.105 |
| 1374–1372 | Aliphatic C–H stretching in $CH_3$ and phen-OH | 1.10 | 1.08 |
| 1266–1252 | C-O stretching vibration in guaiacyl ring | 1.13 | 1.11 |
| 1162* | C-H stretching vibration in syringyl ring | - | 1.08 |
| 1060–1057 | C-O bending vibration in primary alcohol, ether | 1.38 | 1.47 |
| 617* | CH asymmetrical stretching vibration in $CH_3$, $CH_2$, CH | 0.97 | 0.95 |

Relative intensity was calculated as the ratio of absorbance of the band to the absorbance of band at 1510 $cm^{-1}$.
*Bands at 1737 and 1162 $cm^{-1}$ appear only after enzymatic treatment.
doi:10.1371/journal.pone.0072346.t004

## Scanning Electron Microscopy for Fiber Morphology

As can be observed from Figure 5, fiber surface of P-C pulp appears to be smooth. After P-X-L-C treatment the fiber surface became rough and fibrillation appeared on the surface indicating delignification on fiber surface which caused release of fibrils [29]. This is in accordance with the results of FTIR which shows that degradation of lignin occurred on the surface after enzymatic treatment. Other morphological changes like appearance of perforations and cracks probably occurred due to enzymatic hydrolysis by xylanase. These morphological changes increased the surface area of contact between chemicals and pulp fibers and resulted in reduced chemical consumption.

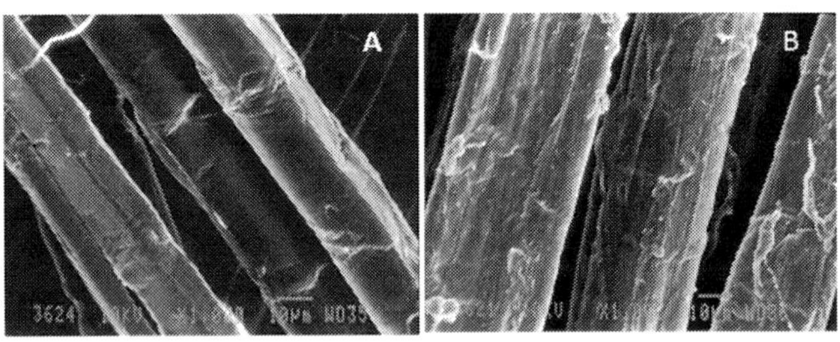


**Figure 5:** Scanning electron micrographs of (A) P-C and (B) P-X-L-C treated ONP pulp.

https://doi.org/10.1371/journal.pone.0072346.g005

## Effect of Sonication (S) and Microwaving (MW) Pretreatments on Deinking of ONP

The parameters for the effect of sonication and microwave pretreatments alone and in combination with P-X-L-C pretreatment of ONP were optimized (Table 5). Maximum brightness of 52.1±0.6 and 52.8±0.41% ISO was obtained when sonication was done at 1% consistency, 230 W, 20 kHz for 15 min and when microwaving was done at 5% consistency, 850 W for 1 min.

**Table 5: Optimization of parameters for sonication (S) and micro**wave (MW) treatment of ONP pulp

https://doi.org/10.1371/journal.pone.0072346.t005

| SONICATION (S) | | | MICROWAVING (MW) | | |
|---|---|---|---|---|---|
| | Brightness (%ISO) | | | Brightness (%ISO) | |
| Time (min) | P-S-C | P-S-X-L-C | Time (sec) | P-MW-C | P-MW-X-L-C |
| 0 | 49±0.45 | 58.65±0.40 | 0 | 49±0.9 | 58.8±0.5 |
| 5 | 50±0.58 | 59.5±0.40 | 30 | 50.1±0.8 | 59.75±0.9 |
| 10 | 51.2±0.8 | 60.35±0.8 | 60 | 52.8±0.41 | **61.73±0.33** |
| 15 | 52.1±0.6 | **61.45±0.75** | 90 | 49.2±0.4 | 57.1±0.44 |
| 20 | 52±0.44 | 61.15±0.39 | 120 | 49.12±0.30 | 53.1±0.39 |

doi:10.1371/journal.pone.0072346.t005

When used in combination with P-X-L-C pretreatment, further 3 (P-S-X-L-C) and 3.5% (P-MW-X-L-C) increase in brightness to 61.45±0.75 and 61.73±0.33% ISO was observed. Sonication causes swelling of fibers thus opening them which in turn increased the surface area of contact between fibers and enzymes or chemicals allowing them to reach protected inertial sites [30]. Sonication has been used previously for deinking of xerox print pulp showing a 10.39% increase in brightness [11]. Microwaving causes steam explosion thereby opening the fibers due to which hemicelluloses and lignin components get exposed to enzymes and chemicals [12]. It is also an efficient, fast, economic, easy and environmental friendly alternative to conventional heating of pulp for enzymatic or chemical bleaching. Microwaving has been explored previously in our lab for delignification of kraft pulp [13]. This study reports for the first time potential of microwave pretreatment for deinking purposes.

## Physical and Chemical Characterization of ONP Pulp (Table 1)

When all the three pretreatments (P-S-MW-X-L-C) were used together, the combination resulted in 62.13±0.5% ISO brightness (28.8% increase), 7.4±0.66 kappa number (22% decrease) and 139.04±22 ppm ERIC (73.9% reduction) as compared to brightness and ERIC of P-C pulps. Substantial reduction in ERIC to this level with the use of a new sequence combining enzymatic and physical methods for efficient deinking of ONP pulp has not been reported before. Though P-S-MW-X-L-C deinked pulp had lower strength properties than P-X-L-C deinked pulp, yet these were better than those of conventional chemically treated pulp. The breaking length of P-X-L-C deinked pulp was 26, 8.6 and 11% higher; burst factor was 0.6, 1.7 and 0.06% higher and tear factor was 12.1, 3.3 and 5.2% higher than P-X-C, P-L-C and P-S-MW-X-L-C deinked pulp. As none of the enzyme preparations had cellulase activity, a 13% increase in viscosity was observed in P-X-L-C treatment. P-L-C treatment caused 12% reduction in kappa number as compared to P-C, indicating the role of laccase in delignification.

## Deinking of other Wastepaper Pulps

Different wastepaper pulps were effectively deinked by using a combination of xylanase and laccase enzymes (Table 2). Maximum increase in brightness was obtained for ONP pulp (21.6%) followed by magazine pulp (8.3%), inkjet print pulp (4.1%), laser print pulp (3.1%) and xerox paper pulp (1.9%). This suggested that X-L system was able to remove both non-impact (newsprint) as well as impact (inkjet, laser and xerox) ink from waste papers. It was more effective in removing non-impact ink as this ink is not fused to the paper and gets easily removed in the deinking process. Impact ink is fused to the paper, hence being non-dispersible is difficult to remove [1]. A 30% reduction in chemical load with laser, magazines, xerox waste papers and 20% with inkjet waste paper was observed (Table 2).

# CONCLUSIONS

Laccase and xylanase effectively deinked wastepaper pulps without mediator supplementation for laccase activity. XL-pretreatment resulted in better optical and strength properties, highly reduced ERIC, higher cellulose crystallinity, more fibrillar surface and formation of lignin-degradation and conjugated carbonyl groups on the surface. A 50% reduction in chemical consumption for ONP pulp was also achieved. Sonication and microwaving contributed further to the improvement in the deinking of pulp. This work reports for the first time a new sequence combining biological and physical methods for effective deinking of old newsprint yielding better quality paper and substantially reduced requirement of chemicals.

## AUTHOR CONTRIBUTIONS

Conceived and designed the experiments: NC PS. Performed the experiments: APV MP VG. Analyzed the data: APV NC PS. Contributed reagents/materials/analysis tools: NC PS. Wrote the paper: APV NC PS.

## REFERENCES

1. Mohandass C, Raghukumar C (2005) Biological deinking of inkjet-printed paper using *Vibrio alginolyticus* and its enzymes. J Ind Microbiol Biotechnol 32: 424–429.
2. Ibarra D, Monte MC, Blanco A, Martinez AT, Martinez MJ (2012) Enzymatic deinking of secondary fibers: cellulases/hemicellulases versus laccase-mediator system. J Ind Microbiol Biotechnol 39: 1–9.
3. Maity C, Ghosh K, Halder SK, Jana A, Adak A, et al. . (2012) Xylanase isozymes from the newly isolated *Bacillus* sp. CKBx1D and optimization of its deinking potentiality. Appl Biochem Biotechnol DOI: 101007/s12010-012-9556-4.
4. Call HP, Strettmatter G (1992) Application of lignolytic enzymes in the paper and pulp industry recent results. Papier J 46: 32–36.
5. Xu Q, Fu Y, Gao Y, Qin M (2009) Performance and efficiency of old newspaper by combining cellulase/hemicellulase with laccase-violuric acid system. Waste management 29: 1486–1490.
6. Virk AP, Capalash N, Sharma P (2011) Use of laccase in pulp and paper industry. Biotechnol Prog 28: 21–32.

7. Singh G, Ahuja N, Batish M, Capalash N, Sharma P (2008) Biobleaching of wheat straw-rich soda pulp with alkalophilic laccase from *gamma-proteobacterium* JB: optimization of process parameters using response surface methodology. Bioresour Technol 99: 7472–7479.
8. Eugenio ME, Santos SM, Carbajo JM, Martín JA, Martín-Sampedro R, et al. (2010) Kraft pulp biobleaching using an extracellular enzymatic fluid produced by *Pycnoporus sanguineus.* . Bioresour Technol 101: 1866–1870.
9. Leduc C, Lanteigne-Roch LM, Daneault C (2011) Use of enzymes in deinked pulp bleaching Cell Chem Technol. 45: 657–663.
10. Xu Q, Fu Y, Gao Y, Qin M, Wang Y, et al. (2011) Fiber surface characterization of old newsprint pulp deinked by combining hemicellulase with laccase-mediator system. Bioresour Technol 102: 6536–6540.
11. Bansal MC, Kumar V, Shilpa K (2011) Combined deinking technology to improve the quality of recycled paper. Appita J 23: 145–148.
12. De la Hoz A, Diaz-Ortiz A, Moreno A (2005) Microwaves in organic synthesis: Thermal and non-thermal microwave effects. Chem Soc Rev 34: 164–178.
13. Woldesenbet F, Virk AP, Gupta N, Sharma P (2012) Effect of microwave irradiation on xylanase production from wheat bran and biobleaching of eucalyptus kraft pulp. Appl Biochem Biotechnol 167: 100–108.
14. Bailey MJ, Biely P, Poutanen K (1992) Laboratory testing method for assay of xylanase activity. J Biotechnol 23: 257–270.
15. Bains J, Capalash N, Sharma P (2003) Laccase from a non melanogenic, alkalotolerant *γ-proteobacterium* JB isolated from industrial waste water drained soil. Biotechnol Letts 25: 1155–1159.
16. Virk AP, Capalash N, Sharma P (2012) An alkalophilic laccase from *Rheinheimera* sp. isolate: Production and biobleaching of kraft pulp. Biotechnol Prog 28: 1426–1431.
17. Degryse E, Glandsdorff N, Picrard A (1978) A comparative analysis of extreme thermophilic bacteria belonging to the genus *Thermus*. Arch Microbiol. 117: 189–196.

18. Miller GL (1959) Use of dinitrosalicylic acid reagent for determination of reducing sugar. Analytical Chem 31: 426–428.
19. Patel AN, Grabski AC, Jeffries TW (1993) Chromophore release from kraft pulp by purified *Streptomyces roseiscleroticus* xylanase. App Microb Biotechnol 39: 405–412.
20. Gupta S, Bhushan B, Hoondal GS (2000) Isolation, purification and characterization of xylanase from Staphylococcus sp SG-13 and its application in biobleaching of kraft pulp. J App Microb 88: 325–334.
21. Heinze T, Liebert T (2001) Unconventional methods in cellulose functionalization. Prog Polym Sci 26: 1689–1762.
22. Oh SY, Yoo DI, Shin Y, Kim HC, Kim HY, et al. (2005) Crystalline structure analysis of cellulose treated with sodium hydroxide and carbon dioxide by means of X-ray diffraction and FTIR spectroscopy. Carbohydr Res 340: 2376–2391.
23. Buta JG, Zardrazil F, Gallettti GC (1989) FT-IR determination of lignin degradation in wheat straw by white rot fungus *Stropharia rugosoannulata* with different oxygen concentrations. J Agr Food Chem 37: 1382–1384.
24. Bajpai P (2004) Biological bleaching of chemical pulps. Crit Rev Biotechnol 24: 1–58.
25. Arias ME, Arenas M, Rodríguez J, Soliveri J, Ball AS, et al. (2003) Kraft pulp biobleaching and mediated oxidation of a nonphenolic substrate by laccase from *Streptomyces cyaneus* CECT 3335. Appl Environ Microbiol 69: 1953–1958.
26. Park S, Baker JO, Himmel ME, Parilla PA, Johnson DK (2010) Cellulose crystallinity index: measurement techniques and their impact on interpreting cellulase performance. Biotechnol Biofuels 24 3–10.
27. Vila C, Barneto AG, Fillat A, Vidal T, Ariza J (2011) Use of thermogravimetric analysis to monitor the effects of natural laccase mediators on flax pulp. Bioresour Technol 102: 6554–6561.
28. Xu Q, Fu Y, Qin M, Quin H (2007) Surface properties of old news print laccase-violuric acid system deinked pulp. Appita J 60: 372–377.

29. Geng X, Li K (2002) Degradation of nonphenolic lignin by the white-rot fungus *Pycnoporus cinnabarinus*. . Appl Microbiol Biotechnol 60: 342–346.

30. Tatsumi D, Higashihara T, Kawamura S, Matsumoto T (2000) Ultrasonic treatment to improve the quality of recycled pulp fiber. Wood sci 46: 405–409.

# CHAPTER 9

# MOLECULAR ENGINEERING OF FUNGAL GH5 AND GH26 BETA-(1,4)-MANNANASES TOWARD IMPROVEMENT OF ENZYME ACTIVITY

**Marie Couturier[1,2], Julia Fe´liu[1,2], Sophie Bozonnet[3,4,5], Alain Roussel[6], Jean-Guy Berrin[1,2]**

[1] INRA, UMR1163, Laboratoire de Biotechnologie des Champignons Filamenteux, Polytech Marseille, Marseille, France

[2] Aix Marseille Universite´, Polytech Marseille, Marseille, France

[3] Universite´ de Toulouse; INSA, UPS, INP; LISBP, Toulouse, France

[4] INRA, UMR792, Inge´nierie des Syste`mes Biologiques et des Proce´de´s, Toulouse, France

[5] CNRS, UMR5504, Toulouse, France

[6] Architecture et Fonction des Macromole´cules Biologiques, UMR7257, CNRS, Aix Marseille Universite´, Marseille, France

## ABSTRACT

Microbial mannanases are biotechnologically important enzymes since they target the hydrolysis of hemicellulosic polysaccharides of softwood biomass into simple molecules like manno-oligosaccharides and mannose. In this study, we have implemented a strategy of molecular engineering in the yeast *Yarrowia lipolytica* to improve the specific activity of two fungal endo-mannanases, *Pa*Man5A and *Pa*Man26A, which belong to the glycoside hydrolase (GH) families GH5 and GH26, respectively. Following random mutagenesis and two steps of high-throughput enzymatic screening, we identified several *Pa*Man5A and *Pa*Man26A mutants that displayed improved kinetic constants for the hydrolysis of galactomannan. Examination of the three-dimensional structures of *Pa*Man5A and *Pa*Man26A revealed which of the mutated residues are potentially important for enzyme function. Among them, the *Pa*Man5A-G311S single mutant, which displayed an impressive 8.2-fold increase in $k_{cat}/K_M$ due to a significant decrease of $K_M$, is located within the core of the enzyme. The *Pa*Man5A-K139R/Y223H double mutant revealed modification of hydrolysis products probably in relation to an amino-acid substitution located nearby one of the positive subsites. The *Pa*Man26A-P140L/D416G double mutant yielded a 30% increase in $k_{cat}/K_M$ compared to the parental enzyme. It displayed a mutation in the linker region (P140L) that may confer more flexibility to the linker and another mutation (D416G) located at the entrance of the catalytic cleft that may promote the entrance of the substrate into the active site. Taken together, these results show that the directed evolution strategy implemented in this study was very pertinent since a straightforward round of random mutagenesis yielded significantly improved variants, in terms of catalytic efiiciency ($k_{cat}/K_M$).

# INTRODUCTION

Hemicellulose is a combined designation of a diverse set of abundant non-crystalline carbohydrate polymers, among which mannans are the major component in softwoods [1] and are also present in certain plant seeds [2]. Mannans comprise molecules constituted either by a backbone of β-1,4-linked D-mannose residues, known as mannan, or by a heterogeneous combination of β-1,4-D-mannose and β-1,4-D-glucose units, termed glucomannan. Both can be decorated with α-1,6-linked galactose side chains, and these polysaccharides are referred to as galactomannan and galactoglucomannan, respectively. According to Salmén [3] softwood glucomannans are incorporated into aggregates of cellulose, i.e., they are arranged in parallel with cellulose fibrils, to which they are tightly connected.

Mannans are hydrolyzed by the coordinated action of several types of glycoside hydrolases (GH) among which endo-β-1,4 mannanases (EC 3.2.1.78) are the key enzymes that depolymerize the mannan backbone. They are encountered in families GH5, GH26 and GH113 in the CAZy database [4], [5]. Beta-mannanases are useful in several industrial processes such as reduction of viscosity of coffee extracts [6] or biobleaching of softwood Kraft pulp [7] and it is now acknowledged that they will become increasingly important for the biorefining of lignocellulose, especially from softwood biomass [8]–[11]. In the biorefinery process, enzymatic hydrolysis of lignocellulosic biomass is one of the major bottlenecks due to the recalcitrance of the plant cell wall and the high cost of enzymes, mainly due to the fact that large amounts are required to breakdown lignocellulose to fermentable sugars [12]–[14].

Despite the fact that mannanases are largely exploited in biotechnological applications, only a few studies have so far reported on the improvement of mannanase properties using molecular engineering [15], [16]. Directed evolution is an important tool for improving critical traits of biocatalysts for industrial applications [17]. Recent advances in mutant library creation and high-throughput screening have greatly facilitated the engineering of biocatalysts but to date only few studies describe improvement of biomass-degrading enzymes using molecular evolution [18]–[20]. The major problem with most directed evolution experiments on biomass-degrading enzymes is the setup of high-throughput assays [17], [21], [22] since it is problematic to measure enzyme activities towards insoluble (hemi) cellulosic substrates.

In this study, a random mutagenesis strategy was used to generate variants of two endo-mannanases from the ascomycete fungus *Podospora anserina* that belong to the glycoside hydrolase (GH) families GH5 and GH26 (*Pa*Man5A and *Pa*Man26A, respectively) in order to improve their activity towards galactomannan. To evaluate the activity of mannanase variants produced in the *Yarrowia lipolytica* expression host, an in-house high-throughput method based on the reducing sugar assay [23] was adapted to assay mannanase activity in liquid culture. The results are interpreted in the lights of the three-dimensional structures of both fungal mannanases [24].

## RESULTS AND DISCUSSION

### *Pa*Man5A and *Pa*Man26A Heterologous Expression in *Yarrowia lipolytica*

The yeast *Y. lipolytica* was chosen as host to perform molecular engineering of *Pa*Man5A and *Pa*Man26A since a very high reproducibility in protein expression level was demonstrated [25]. The two genes encoding *Pa*Man5A and *Pa*Man26A were inserted into the *Y. lipolytica*expression vector in frame with the yeast preprolip2 secretion peptide under the control of the oleic acid-inducible promoter *POX2* (Figure 1). Positive transformants selected on plates containing galactomannan were able to produce functional *Pa*Man5A and *Pa*Man26A enzymes to a level of 10.4±0.2 and 11.2±0.6 $U.ml^{-1}$ in shake flasks, respectively. The culture conditions enabling the secretion of wild-type *Pa*Man5A and *Pa*Man26A were miniaturized in 96-well plates as described in [26] to facilitate the set-up of the high-throughput screening procedure. The mean activities calculated from repeated experiments were 1.78±0.14 and 1.24±0.15 $U.ml^{-1}$ of culture for *Pa*Man5A and *Pa*Man26A, respectively (coefficient of variation [CV] of 7.7 and 12.2%, respectively), which is adequate for excluding false positives variants. The secretion yields of *Pa*Man5A and *Pa*Man26A in deep-well microplates were 6- and 9-fold lower than in shake flask cultures, which is probably due to a better oxygenation of cultures in flasks than in deep-well plates where the ratio volume of culture/volume of container was of 1/2 instead of 1/10. However, levels of mannanase production were in both cases sufficient to deploy a high-throughput screening campaign.

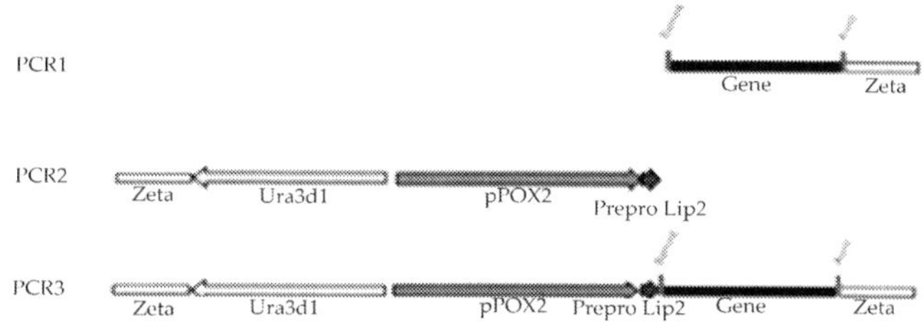


**Figure 1:** Error-prone PCR strategy used in the study.

PCR1: error prone-PCR performed on *paman5a* (HM357135) and *paman26a*(HM357136); PCR2: PCR without mutation performed on Ura3d1 (selection marker), pPOX2 (inducible promotor of acyl-coA oxidase 2) and prepro Lip2 (secretion signal sequence); PCR3: overlapping PCR to reconstruct the entire sequence between zeta platforms. Primers used are listed in Table 3.

https://doi.org/10.1371/journal.pone.0079800.g001

## Construction of First-Round Mutagenesis Libraries and Screening

The objective of the directed evolution approach was to improve enzymatic activities of *Pa*Man5A and *Pa*Man26A, two endo-mannanases from *Podospora anserina.* The error-prone PCR mutagenesis round was carried out using the wild type genes as templates (GenBank HM357135 and HM357136 for *paman5a* and *paman26a*, respectively) and with the conditions appropriate to reach a number of mutations in a range of 2 to 5 mutations per kb. Indeed, a very low number of mutations per kb (i.e., below one mutation per kb) would result in too many active variants with unchanged activity, and a high number of mutations per kb would result in too many inactive clones as reported in [17] (>70% inactive clones for a mutation rate of 10 mutation per kb). In our study, subcloning and sequence analysis of a representative set of mutated genes from each library revealed that the mutation rates were in the desired range (i.e. 2.6 and 4.5 mutations per kb for *Pa*Man5A and *Pa*Man26A, respectively). Figure 2 shows a summary of the libraries screening and the number of variants selected at each step. *Y. lipolytica* transformation yielded about 5,200 and 5,600 clones for *Pa*Man5A and *Pa*Man26A, respectively. As reported in other studies [20], [27], [28], 5,000 variants was considered as a reasonable size of library to identify mutants displaying improved characteristics. The first step of the screening consisted in the selection of active clones on solid medium containing AZO-dyed galactomannan. Since we were looking for an activity improvement that was expected to be rather modest, halo-producing clones could not be screened only by comparing the size of halos but required a second step of screening. Culture-based screening was performed using our in-house robotic platform with about 2,000 and 1,500 mutants for *Pa*Man5A and *Pa*Man26A, respectively (Figure 2). Since the screening capability of this high-throughput system is 15 plates per day, ~1,400 variants could be analysed in one day and the defined screening job (3,500 variants) was finished within 3 days. Among the best-performing clones that were further tested to confirm the enhancement of activity in liquid cultures, we finally selected four mutants of *Pa*Man5A and one mutant of *Pa*Man26A that displayed improved activity beyond the CV of wild-type mannanases (7.7 and 12.2%, respectively). The *Y. lipolytica*strain expressing the best *Pa*Man26A variant displayed an increase of activity towards galactomannan of 147% compared to the strain expressing wild-type *Pa*Man26A that corresponded to 12 CV. The *Y. lipolytica* strains expressing selected *Pa*Man5A variants displayed also increased activity between 8 and 46% compared to the strain expressing wild-type *Pa*Man5A that corresponded to 1.1 to 6 CV. Each of the mannanase-mutant genes was amplified from genomic DNA, subcloned and sequenced. As a result, one single (*Pa*Man5A-G311S), two double (*Pa*Man5A-K139R/Y223H, *Pa*Man26A-P140L/D416G) and two triple mutants (*Pa*Man5A-V256L/G276V/Q316H, *Pa*Man5A-W36R/I195T/V256A) were created. Results are summarized in Table 1.

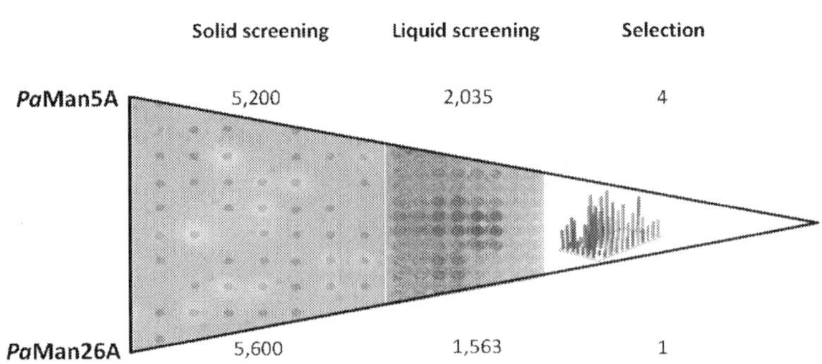


**Figure 2:** Screening strategy and mutant selection.

The number of variants screened at each step is indicated at the top (*Pa*Man5A) and at the bottom (*Pa*Man26A) of the diagram.

https://doi.org/10.1371/journal.pone.0079800.g002

**Table 1:** Mannanase activity of selected *Y. lipolytica* **variants**

https://doi.org/10.1371/journal.pone.0079800.t001

| Enzyme | Activity ($U.ml^{-1}$) | CV (%) | Activity Improvement (%) |
|---|---|---|---|
| *Pa*Man26A wt | 1.24±0.15 | 12.2 | - |
| *Pa*Man26A-P140L/D416G | - | - | 147 |
| *Pa*Man5A wt | 1.78±0.14 | 7.7 | - |
| *Pa*Man5A-V256L/G276V/Q316H | - | - | 46 |
| *Pa*Man5A-W36R/I195T/V256A | - | - | 9 |
| *Pa*Man5A-K139R/Y223H | - | - | 20 |
| *Pa*Man5A-G311S | - | - | 11 |

The mannanase activity was measured at 40°C in sodium acetate buffer 50 mM, pH 5.2 using 1% (w/v) galactomannan. The coefficient of variation (CV) was defined as the ratio of the standard deviation to the mean and was calculated for each of the wild-type enzymes. wt, wild type.

doi:10.1371/journal.pone.0079800.t001

## Mutant production in *P. pastoris* and Biochemical Characterization

For mutant enzymes production, we used the *P. pastoris* expression system because (i) it yields higher expression levels than *Y. lipolytica* [29] and (ii) wild-type *Pa*Man5A and *Pa*Man26A were already successfully expressed to high yields in *P. pastoris* [10]. The ratio of mannanase production yields obtained in the culture medium with *Y. lipolytica* and *P. pastoris*was roughly 1:10. All of the mutant enzymes were successfully produced in *P. pastoris* with yields of approximately 1 g per liter of culture and purified taking advantage of the $(His)_6$-tag. SDS-PAGE analysis of the purified mutants compared to the wild type enzymes revealed similar apparent molecular masses (data not shown). Isoelectrofocusing analysis revealed that all of the mutants displayed similar pIs compared to wild-type enzymes except for the *Pa*Man5A-W36R/I195T/V256A mutant that exhibited an increase of 0.3 unit. This increase is likely to be due to the incorporation of an arginine residue. Circular dichroism analysis of all selected mutants was also carried out to check the relative content of secondary structures in each mutant. *Pa*Man26A and *Pa*Man5A variants displayed the same profile compared to parental enzymes, suggesting that the folding of mutants were similar to the parental enzymes (data not shown). Regarding the pH and temperature profiles, no significant difference was observed between variants and parental enzymes (data not shown).

## Kinetic Parameters of Mutants toward Galactomannan

Each mutant was characterized using High Performance Anion Exchange Chromatography – Pulsed Amperometric Detection (HPAEC-PAD) to assess its ability to hydrolyze galactomannan and manno-oligosaccharides: mannopentaose ($M_5$) and mannohexaose ($M_6$) (Table 2). Regarding the $M_5$ and $M_6$ hydrolysis, *Pa*Man26A-P140L/D416G exhibited 100% and 30% increase of $k_{cat}/K_M$ toward $M_5$ and $M_6$, respectively. *Pa*Man5A-V256L/G276V/Q316H and *Pa*Man5A-W36R/I195T/V256A exhibited same hydrolysis profile as *Pa*Man5A at 10 min or 20 min manno-oligosaccharide hydrolysis (data not shown) and therefore $k_{cat}/K_M$ were considered unchanged and not determined. *Pa*Man5A-K139R/Y223H revealed a decrease of $k_{cat}/K_M$ of about 35% and 11% respectively. *Pa*Man5A-G311S displayed an increased $k_{cat}/K_M$ of about 37% and 12% towards $M_5$ and $M_6$, respectively.

**Table 2:** Kinetic constants of wild-type enzymes and selected variants toward galactomannan, mannohexaose ($M_6$) and mannopentaose ($M_5$)

https://doi.org/10.1371/journal.pone.0079800.t002

| | Galactomannan | | | $M_6$ | $M_5$ |
|---|---|---|---|---|---|
| Enzyme | $K_M$ (mg.ml$^{-1}$) | $k_{cat}$ (min$^{-1}$) | $k_{cat}/K_M$ (mg$^{-1}$.ml.min$^{-1}$) | $k_{cat}/K_M$ (mg$^{-1}$.ml.min$^{-1}$) | |
| *Pa*Man26A wt | 2.4±0.3 | 3356±159 | 1413±155 | $7.6\times10^5$ | $3.8\times10^5$ |
| 26-P140L/D416G* | 2.1±0.2 | 3860±80 | 1849±148 | $1.0\times10^6$ | $7.8\times10^5$ |
| *Pa*Man5A wt | 11.5±1.5 | 1674±138 | 147±19 | $3.0\times10^6$ | $9.2\times10^5$ |
| 5-V256L/G276V/Q316H | 7.8±1.6 | 1493±167 | 197±38 | ND | ND |
| 5-W36R/I195T/V256A* | 16.2±2.5 | 4199±438 | 263±41 | ND | ND |
| 5-K139R/Y223H* | 6.7±0.5 | 1655±60 | 248±17 | $2.7\times10^6$ | $6.8\times10^5$ |
| 5-G311S* | 1.5±0.4 | 1781±134 | 1247±292 | $3.4\times10^6$ | $1.2\times10^6$ |

The kinetic parameters were determined at 40°C in sodium acetate buffer 50 mM, pH 5.2 as described in the Methods section. Paired t test was used to compare the kinetic parameters of mutants versus native enzyme. The difference was considered statistically significant when $p<0.05$ (*). wt, wild type.ND: not determined.
doi:10.1371/journal.pone.0079800.t002

Regarding the $K_M$ apparent values, all of the mutants displayed improved apparent affinity for galactomannan. Although turn-over numbers of almost all of the mutant enzymes were similar to wild-type enzymes, the *Pa*Man5A-W36R/I195T/V256A mutant displayed a 2.5-fold increase of turn-over toward galactomannan resulting in an overall catalytic efficiency improved by 1.8-fold. In terms of catalytic efficiency, the best-performing mutant was the *Pa*Man5A-G311S mutant for which the single amino-acid substitution led to a 8.2-fold increase in $k_{cat}/K_M$ due to a drastic improvement of $K_M$ whereas the $k_{cat}$ remained unchanged. Other *Pa*Man5A mutants exhibited also to a lower extent increased $k_{cat}/K_M$ towards galactomannan, i.e., between 32 and 79% improvement. The *Pa*Man26A-P140L/D416G mutant, which is the unique mutant selected from the 5,600 *Pa*Man26A variants screened, displayed an improved $k_{cat}/K_M$ of approximately 30%. All of the mutants were subjected to a paired t-test in comparison with the corresponding native enzyme. Three out of four *Pa*Man5A mutants and the *Pa*Man26A mutant revealed *p*-values below 0.05 for galactomannan hydrolysis and were therefore considered statistically different from their native counterpart (Table 2).

Only sparse studies have previously led to the identification of mutants displaying such increase in catalytic efficiency. For example, several endoglucanase mutants with $k_{cat}/K_M$ improvement between 15 and 80% were generated after three rounds of mutagenesis [26]. Song *et al* [20] identified several xylanase mutants with $k_{cat}/K_M$ towards xylan increased in a range of 20 to 50% after several rounds of evolution.

## Structure-Function Relationships of Identified Variants

Examination of the recently solved three-dimensional structures of *Pa*Man5A and *Pa*Man26A [24] revealed that some of the mutated residues are potentially important for enzyme function.

The localization of all of the mutated residues of the *Pa*Man5A mutants revealed that five mutated residues (W36R, K139R, Y223H, V256A/L, and Q316H) out of eight are located within the active site cleft and three mutations are located inside the enzyme core (I195T, G276V and G311S). Three mutations were identified in the *Pa*Man5A-W36R/I195T/V256A mutant, among which W36 and V256 that are located close to the catalytic site (Figure 3A) and I195T is located in the enzyme core (not shown). The W36 residue side chain lies at the bottom of the active site crevice and may contribute to the −4 subsite. Interestingly, the Val256 does not seem to be directly involved in the subsite organization of *Pa*Man5A, but mutation of this residue were found in two distinct variants (*Pa*Man5A-V256L/G276V/Q316H and *Pa*Man5A-W36R/I195T/V256A), indicating a possible hot spot for enzyme improvement to further explore. Concerning the *Pa*Man5A-K139R/Y223H mutant, analysis of the manno-oligosaccharides produced upon hydrolysis of $M_5$ and $M_6$ revealed a potential modification of substrate binding (Figure 4). Indeed, the wild type *Pa*Man5A is known to produce mainly $M_3$ from $M_6$, with small amounts of $M_2$ and $M_4$ [24]. In the case of *Pa*Man5A-K139R/Y223H, equimolar amount of $M_2$, $M_3$ and $M_4$ were quantified (Figure 4), suggesting a displacement of the substrate binding from −3 to +3 subsites to −2 +4 subsites or −4 +2 subsites (Figure 3). Within the two mutations of mutant *Pa*Man5A-K139R/Y223H, it is interesting to note that the Y223 residue is located nearby the +3 subsite. The substitution of the aromatic ring, involved in stacking interaction with the sugar, may contribute to the modification of substrate binding at this position (Figure 3). The single mutation of *Pa*Man5A-G311S is located in the core of the enzyme within the start of β8-strand (Figure 3B). The G311 residue is strictly conserved among all of the GH5 mannanases of known structure. The G311 residue is conserved among GH5 mannanases within the start of the β8-strand and it should be noted that the last residue of this strand is the W315 residue (Figure 3B), which shapes the −2 subsite. The G311 residue is in close vicinity with a compact hydrophobic core made of residues F20, I60, I115, F170, F244, L280 and L313 (Figure 3C). It is difficult to determine the effect of this single substitution from a structural point of view, although we can hypothesize that the hydroxyl group of the serine residue may contribute to the modification of the β8-strand by interacting with the surrounding residues although circular dichroïsm analysis of the *Pa*Man5A-G311S mutant revealed a similar profile compared to the parental enzyme, suggesting that its folding was similar to the parental enzyme (data not shown). Therefore the single mutation *Pa*Man5A-G311S does not seem to modify the overall architecture of the enzyme but a slight movement of the β8-strand could move the W315 residue at the surface of the enzyme and explain the decreased $K_M$.

A. Surface view of the catalytic cleft of *Pa*Man5A with mannotriose modelled in the −2 and −3 subsites and mannobiose modelled in the +1 and +2 subsites. The structures of GH5 from *T. reesei* and *T. fusca* in complex with mannobiose and mannotriose, respectively, were superimposed on the top of the structure of *Pa*Man5A to map the substrate-binding subsites. The two catalytic glutamate residues, E177 and E283, are coloured in red. The substituted amino-acids are labelled and coloured in yellow. B. Structural based sequence alignment of the region around position 311 (according to *Pa*Man5A numbering) from *Podospora anserina* (*Pa*Man5A), *Aplysia kurodai* (*Ak*Man, PDB 3VUP), *Mytilus edulis* (*Me*Man5A, PDB 2C0H), *Cellvibrio mixtus* (*Cm*Man5A, PDB 1UUQ), *Trichoderma reesei* (*Tr*Man5A, PDB 1QNR), *Lycopersicon esculentum*(*Le*Man4A, PDB 1RH9) and *Thermomonospora fusca* (*Tf*Man5, PDB 2MAN). Secondary structure elements, α-helix α7 and β-strand β8, are indicated below the sequences as a cylinder and an arrow, respectively. Strictly conserved residues, G311 and W315 (according to *Pa*Man5A numbering), are shown with a yellow and a grey background, respectively. C. Surface view of *Pa*Man5A rotated of about 90° along the horizontal axis. The front clipping plane has been moved in order to visualize the location of G311 inside the molecule. The zoom shows a compact hydrophobic core in the vicinity of G311.

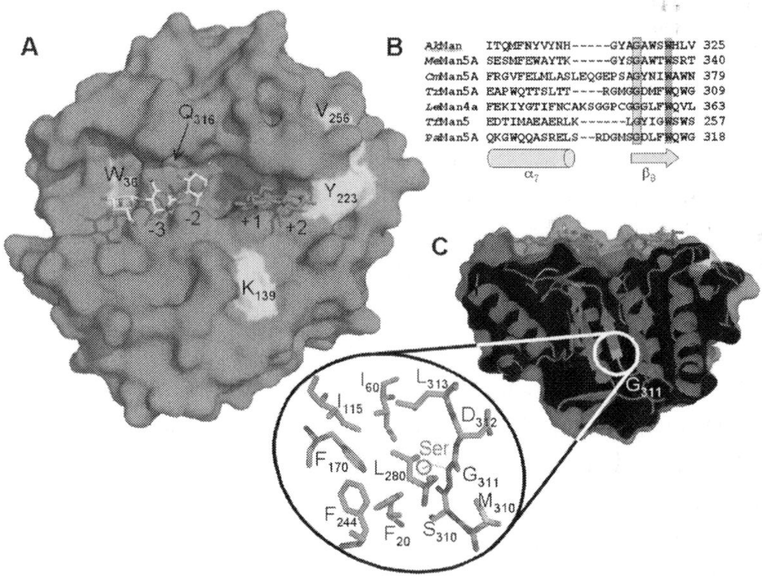


**Figure 3:** Structural view of *Pa*Man5A (PDB 3ZIZ) exhibiting substituted amino-acids.

https://doi.org/10.1371/journal.pone.0079800.g003

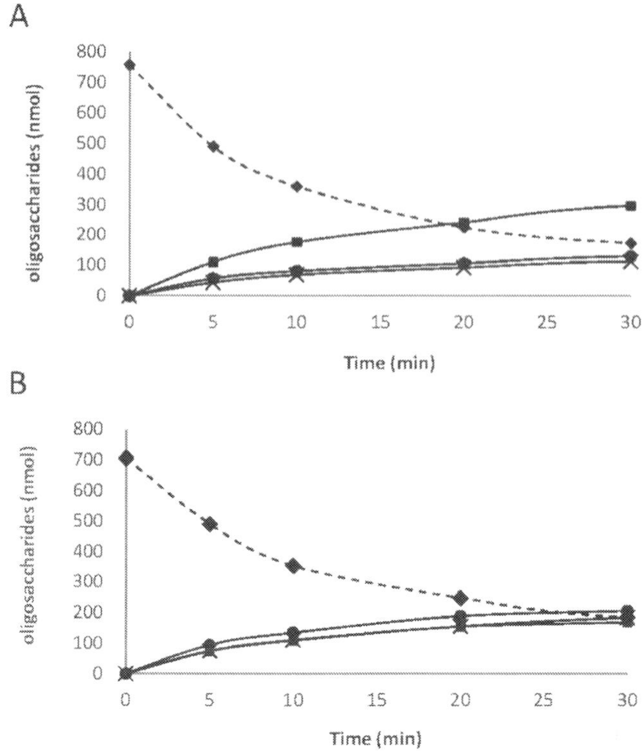


**Figure 4: Progress curves of the** manno-oligosaccharides generated by the wild-type *Pa*Man5A and the *Pa*Man5A-K139R/Y223H variant upon hydrolysis of mannohexaose.

18.2 nM of the wild-type *Pa*Man5A (A) and the *Pa*Man5A-K139R/Y223H variant (B) were incubated with 1 mM of mannohexaose in acetate buffer pH 5.2 at 40°C. The amount of each manno-oligosaccharide, i.e., mannobiose (full circles), mannotriose (full squares), mannotetraose (crosses), and mannohexaose (full diamonds), is indicated during the course of the reaction.

https://doi.org/10.1371/journal.pone.0079800.g004

One of the most remarkable findings of this study is the activity increase of the *Pa*Man26A-P140L/D416G mutant that contained only two mutations, i.e., one in the linker region (P140L) and one at the entrance of the active site (D416G) (Figure 5). The region from residue R133 to residue N141, which may be considered as the end of the linker, is tightly bound to the catalytic domain thanks to a dense network of hydrogen bonds and hydrophobic interactions [24]. Moreover, the *Pa*Man26A linker sequence contains four proline residues (P132, P134, P135 and P140) out of 12 residues that may confer rigidity to the modular enzyme [24]. In the *Pa*Man26A-P140L/D416G mutant, the P140 residue was substituted by a leucine residue, probably resulting in a decrease of linker rigidity that could partially explain the improved $k_{cat}/K_M$. The D416 amino-acid substitution is located at the edge of the catalytic cleft (Figure 5). The favourable mutation consisted in the removal of the carboxylic group since the D416 amino-acid was substituted by a glycine residue. We suggest that the lack of the carboxylic acid side chain may promote the entrance of the substrate into the active site.

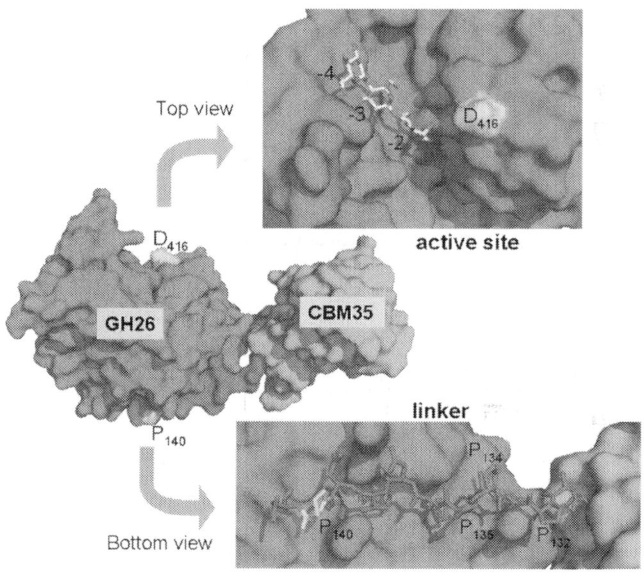


**Figure 5:** Structural view of *Pa*Man26A (PDB 3ZM8) exhibiting substituted amino-acids.

The central panel shows a surface view of the entire *Pa*Man26A structure, which is composed of a carbohydrate binding module (CBM) belonging to the CBM35 family in cyan, a linker in violet and a catalytic domain belonging to the GH26 family in green. The two catalytic glutamate residues, E300 and E390, are coloured in red. The two substituted amino-acids, P140 and D416, are labelled and coloured in yellow. The top view represents the surface view of the catalytic cleft of

*Pa*Man26A rotated about 90° along the horizontal axis with mannotriose modelled into the −2 to −4 subsites. The structure of GH26 from *C. fimi* in complex with mannotriose was superimposed on the top of the structure of *Pa*Man26A to map the substrate-binding subsites. The bottom view displays the *Pa*Man26A linker (from residue 131 to residue 141) in stick representation. The molecule has been rotated of about 90° along the horizontal axis and in the opposite direction compared to the top view. The proline residues of the linker are labelled.

https://doi.org/10.1371/journal.pone.0079800.g005

# CONCLUSIONS

Activity improvement of fungal carbohydrate-active enzymes has typically met many obstacles, owing primarily to the lack of efficient expression systems and high-throughput screening methodologies. In this study, molecular engineering of two fungal mannanases was undertaken using random mutagenesis in the yeast *Y. lipolytica*. All of the selected mutants highlighted improved characteristics when compared to wild-type enzymes, thus validating the approach of mutagenesis and screening that were employed. Examination of the three-dimensional structures of *Pa*Man5A and *Pa*Man26A revealed several hot spots (i) around the active site of both mannanases, (ii) in the core region of *Pa*Man5A and (iii) in the linker region of *Pa*Man26A. The directed evolution strategy implemented in this study was very successful since a straightforward round of random mutagenesis yielded significantly improved variants that would have been difficult to predict using rational site-directed mutagenesis and structure-guided design.

# METHODS

## Plasmids, Strains and Culture Conditions

The growth media and culture conditions for the *Y. lipolytica* JMY1212 strain (Ura-, Leu+, ΔAEP, Suc+) have been previously described [30]. Briefly, the *Y. lipolytica* JMY1212 strain was cultured in YPD (10 g/l yeast extract, 20 g/l peptone, 20 g/l glucose) at 28°C, 130 rpm in baffled flasks or in Petri dishes containing YPD supplemented with agar (15 g/l). The plasmid used for expression in *Y. lipolytica* was JMP61 that displayed the *POX2* promoter for induction by oleic acid [31] and the Zeta recombination platform for controlled monocopy integration in *Y. lipolytica* genome. Subcloning of parental genes was performed using the TOPO TA kit (Invitrogen, Cergy-Pontoise, France) and TOP10 *E. coli* competent cells (Invitrogen).

## Construction of JMP61-*paman5a* and JMP61-*paman26a* Plasmids and Yeast Transformation

*Pa*Man5A and *Pa*Man26A genes (*paman5a* [HM357135] and *paman26a* [HM357136], respectively) were amplified with primers listed in Table 3, starting from the pPICZαA-*paman5a*and the pPICZαC-*paman26a* plasmids [10]. Insertion of both genes in the JMP61 expression vector was performed using *Bam*HI and *Avr*II restriction sites. *Y. lipolytica* Zeta competent cells were prepared using the lithium acetate method as described in le Dall *et al* [32].

**Table 3:** List of primers used in the study

https://doi.org/10.1371/journal.pone.0079800.t003

| Gene | PCR step | Primer name | Sequence (5′->3′) |
|---|---|---|---|
| *paman5a* | integration in JMP61 | GH5JMP61F 5′ | TTTGGATCCCTCCCCCAAGCACAA |
| | | GH5JMP61R 3′ | TTTCCTAGGCTACGCCGGGAGAGCATT |
| | mutagenesis PCR (PCR1) | JMP61EvDF | GCAGAAGCGATTCGGATCC |
| | | PCR2r | GGAGTTCTTCGCCCACCC |
| | zeta platform PCR (PCR2) | PCR1d | GATCCCCACCGGAATTGC |
| | | GH5EvDRev | GGCTGCTCCTCCACC |
| | integration in pPICZαA | GH5-InFuFOR | GGCTGAAGCTGAATTCCTCCCCCAAGCACAAGGT |
| | | GH5-InFuR | GAGTTTTTGTTCTAGACCCCGCCGGGAGAGCATT |
| *paman26a* | integration in JMP61 | JMP61-CBMGH26-F | TTTGGATCCAAGCCTTGTAAGCC |
| | | GH26-JMP61-Rev | TTTCCTAGGCTAACTCCTCCACCCCTGAAT |
| | mutagenesis PCR (PCR1) | PCR1dtmutCBM-F | TTTCCAACCTCAACAACCCCAAC |
| | | PCR2r | GGAGTTCTTCGCCCACCC |
| | zeta platform PCR (PCR2) | PCR1d | GATCCCCACCGGAATTGC |
| | | GH26EvDRev | GGAGTAGAGCTTCTT |
| | integration in pPICZαA | GH26CBM-InFuFOR | GGCTGAAGCTGAATTCAAGCCTTGTAAGCCTCGT |
| | | GH26InFuR | GAGTTTTTGTTCTAGACCACTCCTCCACCCCTGAAT |
| common | overlap PCR (PCR 3) | PCR1dL | CCGCTGTCGGGAACCGCGTTCAGGTGGAACAGG |
| | | PCR2rL | CCGCACTGAGGGCTTTGTGAGGAGGTAACGCCG |

doi:10.1371/journal.pone.0079800.t003

Competent cells were immediately transformed with 500 ng *Not*I-linearized recombinant plasmid combined with 25 μg of salmon sperm DNA by heat shock at 39°C for 10 minutes and immediately recovered in 1.2 ml of 100 mM lithium acetate. Transformants were plated on

YNB agar (1.7 g/l YNB, 10 g/l glucose, 5 g/l ammonium chloride, 2 g/l casamino acids, in 50 mM sodium–potassium phosphate buffer, pH 6.8, 17 g/l agar) and incubated at 28°C for 48 hours.

## Construction of Mutagenesis Libraries by Error-Prone PCR

Three PCR were carried out for each gene as shown in Figure 1 with primers listed in Table 3. Mutagenic PCR (PCR1) were performed using Genemorph II Random Mutagenesis Kit (Stratagene, La Jolla, CA) following the manufacturer›s instructions using primers JMP61EvDF and PCR2r for *paman5a* and PCR1dtmutCBM-F and PCR2r for *paman26a*. The PCR contained 500 ng of parental gene, 200 μM dNTP, 0.25 μM primers and 2.5 units of Mutazyme. The PCR were made up to 50 μl and incubated at 95°C for 5 min and then at 95°C for 30 sec, 50°C for 30 sec, 72°C for 1 min and 30 sec for 40 cycles followed by 5 min at 72°C. PCR2 was carried out using HiFi polymerase (Invitrogen) to amplify the zeta platform without introducing any mutations. Primers used were PCR1d and GH5EvDRev for *paman5a* and PCR1d and GH26EvDRev for *paman26a*. PCR products were gel-extracted using Gel-Extraction kit (Qiagen, Courtaboeuf, France) and a third overlap PCR (PCR3) was subsequently carried out using gel-extracted products of PCR1 and PCR2 to rebuild the whole zeta fragment containing mutant genes. The primers used for PCR3 were PCR1dL and PCR2rL for the two constructions. The overlap PCR was carried out using i-Star Max II DNA polymerase (Intron Biotechnology, Boca Raton, FL).

PCR3 products were subsequently transformed into *Y. lipolytica* as described above and plated onto YNB in QTrays plates (Corning Corp, NY, USA) at a density of about 500 colonies par plate and incubated for 48 hours at 28°C.

## Screening of Mutagenesis Libraries

### *Agar plate-based screening*

The Ura-positive *Y. lipolytica* transformants obtained from mutagenesis libraries were subsequently gridded on YNB agar medium containing oleic acid (1% v/v, Sigma) and Azo-galactomannan (0.2% w/v,

Megazyme) using a QPixII colony picker (Genetix, Molecular Devices, Sunnyvale, CA, USA). The plates were further incubated at 28°C for 48 hours. Mannanase activity was visualized as clear halos around colonies within a blue background.

### *Liquid culture-based screening*

Mannanase-positive colonies were picked on OmniTrays (Nunc, Thermo Fischer Scientific, Courtaboeuf, France) containing YPD agar and incubated for 48 hours at 328°C. Further screening was performed as described in [26]. Briefly, 96-well preculture plates containing 200 µl YPD were inoculated by individual colonies. The precultures were incubated at 800 rpm and 28°C overnight in a Microtron incubator (Infors HT, Switzerland). For expression of recombinant genes, 20 µl of each preculture were transferred in 1 ml of YTO medium (10 g/l yeast extract, 20 g/l% w/v tryptone, 2% oleic acid, in 50 mM phosphate buffer, pH 6.8) in 96-deep-well plates. Cultures were further incubated at 28°C with shaking at 800 rpm. After 4 days induction, the supernatants containing mutant enzymes were recovered by centrifugation (10 min, 3500 rpm) and endo-mannanase activity towards galactomannan was determined from DNS assay as described before [23]. Briefly, 10 µl of culture supernatant were incubated with 190 µl of 1% (w/v) galactomannan in 50 mM sodium phosphate buffer (pH 5) at 40°C. 80 µl of reaction mixture was recovered and reaction was terminated by addition of same volume of dinitrosalicylic acid reagent at 1% (w/v) in 96 well PCR plates. Samples were heated at 95°C for 10 min and $DO_{540}$ was measured relative to a mannose standard curve (1 to 20 mM).

## Wild-Type and Mutant Enzymes Large-Scale Production

For heterologous production of *Pa*Man5A and *Pa*Man26A mutant proteins, the selected genes were amplified from genomic DNA using GH5-InFuFOR and GH5-InFuR for *Pa*Man5A and GH26CBM-InFuFOR and GH26-InFuR for *Pa*Man26A, respectively, listed in Table 3 and transferred into the pPICZαA plasmid using InFusion kit (Clontech, Takara). Resulting plasmids were transformed into *P. pastoris* and protein productions and purification were carried out as described before [10].

## Biochemical and Biophysical Characterization

Protein concentration was determined by using the Bio-Rad protein assay kit with bovine serum albumin as standard (Bio-Rad, Marnes-la Coquette, France) and UV absorbance at 280 nm. SDS-PAGE was performed in 10% (w/v) polyacrylamide gel (Bio-Rad) using a Pharmacia LMW electrophoresis calibration kit (GE Healthcare, Buc, France). Native IEF was carried out at 4°C in the Bio-Rad gel system, using pI standards ranging from 4.45 to 8.2. IEF gel was coloured with IEF staining solution (0.04% Coomassie Blue R250, 0.05% Cocrein Scarlett, 10% acetic acid, 27% isopropanol).

## Characterization and Kinetic Properties of Individual Enzymes

Determination of kinetic parameters on galactomannan was performed using DNS activity assay. Unless otherwise indicated, assay mixtures contained substrate and suitably diluted enzyme in sodium acetate buffer 50 mM, pH 5. Briefly, 1 μg of enzyme was mixed with 190 μl of galactomannan (Megazyme International, Wicklow, Ireland) using a range of substrate concentration from 1 to 20 $mg.ml^{-1}$ (eight concentrations) and incubated at 40°C for 5 minutes. Reactions were performed in triplicate independent experiments. The reaction was stopped by the addition of 300 μl of 1% DNS reagent and samples were heated at 95°C for 10 minutes. $DO_{540}$ was measured relative to a mannose standard curve (0 to 20 mM). One unit of endo-mannanase activity was defined as the amount of protein that released 1 μmol of sugar monomers per min. The kinetic parameters were estimated using weighted nonlinear squares regression analysis with the Grafit software (Erithacus Software, Horley, UK).

## Analysis of Initial Sugar release by HPAEC-PAD and Kinetic Parameters Measurement

Monosaccharide and oligosaccharides generated after hydrolysis of manno-oligosaccharides ($M_5$ and $M_6$, Megazyme) were analysed using HPAEC-PAD. 10 μl of suitably-diluted enzyme were incubated at 40°C for various time lengths with 190 μl of 1 mM substrate in 50 mM acetate buffer pH 5.2. At each time point, 10 μl of reaction mixture were recovered and the reaction was terminated by the addition of 90 μl of 180 mM NaOH. For HPAEC analysis, 10 µl were injected and elution was carried out as described in [10]. Calibration curves were plotted using β-1,4-manno-oligosaccharides as standards from which response factors were calculated (Chromeleon program, Dionex) and used to estimate the amount of products released in test incubations. All the assays were carried out at least in duplicates. The specificity constants were calculated using the Matsui equation [24], [33].

## ACKNOWLEDGMENTS

The authors want to thank M. Haon and S. Grisel for their assistance in the purification of mutants. C. Montanier and C. Dumon (LISBP, INSA, Toulouse, France), P.M. Coutinho (AFMB, CNRS, Marseille, France), N. Lopes Ferreira (IFPEN, Rueil-Malmaison, France) and D. Navarro (INRA, Marseille, France) are acknowledged for helpful discussions and C.B. Faulds for proofreading of the manuscript. The plasmids and strains for *Yarrowia* expression experiments were generously provided by JM Nicaud (MICALIS, INRA-AgroParisTech, France) and A Marty (LISBP, INSA, Toulouse, France). The high-throughput screening work was carried out at the Laboratory for BioSystems & Process Engineering (Toulouse, France) with the equipments of the ICEO facility, dedicated to the screening and the discovery of new and original enzymes. This work was carried out as part of the FUTUROL PROJECT and the authors want to thank OSEO Innovation for its financial support.

## AUTHOR CONTRIBUTIONS

Conceived and designed the experiments: MC JGB. Performed the experiments: MC JF SB. Analyzed the data: MC AR JGB. Contributed reagents/materials/analysis tools: JGB. Wrote the paper: MC JGB.

## REFERENCES

1. 1.Wiedenhoeft AC, Miller RB (2005) Structure and Function of Wood. In: Rowell RM, editor. Handbook of wood chemistry and wood composites. CRC Press. 9–33.
2. 2.Scheller HV, Ulvskov P (2010) Hemicelluloses. Annu Rev Plant Biol 61: 263–289.

3. Salmén L (2004) Micromechanical understanding of the cell-wall structure. C R Biol 327(9–10): 873–880.
4. Cantarel BL, Coutinho PM, Rancurel C, Bernard T, Lombard V, et al. (2009) The Carbohydrate-Active enZyme database (CAZy): an expert resource for glycogenomics. Nucleic Acids Res 37: 233–238.
5. The Carbohydrate-Active enZymes Database [www. cazy. org].
6. Sachslehner A, Foidl G, Foidl N, Gübitz G, Haltrich D (2000) Hydrolysis of isolated coffee mannan and coffee extract by mannanases of *Sclerotium rolfsii*. J Biotechnol 80: 127–134.
7. Montiel MD, Rodriguez J, Perez-Leblic MI, Hernandez M, Arias ME, et al. (1999) Screening of mannanases in actinomycetes and their potential application in the bleaching of pine Kraft pulps. Appl Microbiol Biotechnol 52: 240–245.
8. Do BC, Dang TT, Berrin JG, Haltrich D, To KA, et al. (2009) Cloning, expression in *Pichia pastoris*, and characterization of a thermostable GH5 mannan endo-1,4-β-mannosidase from *Aspergillus niger* BK01. Microb Cell Fact 8: 59.
9. Pham TA, Berrin JG, Record E, To KA, Sigoillot JC (2010) Hydrolysis of softwood by *Aspergillus* mannanase: role of a carbohydrate-binding module. J Biotechnol 148: 163–170.
10. Couturier M, Haon M, Coutinho P, Henrissat B, Lesage-Meessen, et al (2011) *Podospora anserina* hemicellulases potentiate the *Trichoderma reesei* secretome for saccharification of lignocellulosic biomass. Appl Environ Microbiol 77(1): 237–246.
11. Chauhan PS, Puri N, Sharma P, Gupta N (2012) Mannanases: microbial sources, production, properties and potential biotechnological applications. Appl Microbiol Biotechnol 93(5): 1817–1830.
12. Merino ST, Cherry J (2007) Progress and challenges in enzyme development for biomass utilization. Adv Biochem Eng Biotechnol 108: 95–120.
13. Himmel EM, Picataggio SK (2008) Biomass Recalcitrance: Deconstructing the Plant Cell Wall for Bioenergy. In: Himmel EM, editor. Biomass Recalcitrance: Deconstructing the Plant Cell Wall for Bioenergy. Blackwell Publishing Ltd. 1–6.

14. Margeot A, Hahn-Hagerdal B, Edlund M, Slade R, Monot F (2009) New improvements for lignocellulosic ethanol. Curr Opin Biotechnol 20: 1–9.
15. Tailford LE, Ducros VM, Flint JE, Roberts SM, Morland C, et al. (2009) Understanding how diverse beta-mannanases recognize heterogeneous substrates. Biochemistry 48(29): 7009–7018.
16. Hekmat O, Lo Leggio L, Rosengren A, Kamarauskaite J, Kolenova K, et al. (2010) Rational engineering of mannosyl binding in the distal glycone subsites of *Cellulomonas fimi* endo-beta-1,4-mannanase: mannosyl binding promoted at subsite −2 and demoted at subsite −3. Biochemistry 49(23): 4884–4896.
17. Wang XJ, Peng YJ, Zhang LQ, Li AN, Li DC (2012) Directed evolution and structural prediction of cellobiohydrolase II from the thermophilic fungus *Chaetomium thermophilum*. Appl Microbiol Biotechnol 95(6): 1469–1478.
18. Heinzelman P, Snow CD, Wu I, Nguyen C, Villalobos A, et al. (2009) A family of thermostable fungal cellulases created by structure-guided recombination. *Proc Natl Acad Sci USA* 106(14): 5610–5615.
19. García-Ruiz E, Maté D, Ballesteros A, Martinez AT, Alcalde M (2010) Evolving thermostability in mutant libraries of ligninolytic oxidoreductases expressed in yeast. Microb Cell Fact 9: 17.
20. Song L, Siguier B, Dumon C, Bozonnet S, O›Donohue MJ (2012) Engineering better biomass-degrading ability into a GH11 xylanase using a directed evolution strategy. Biotechnol Biofuels 5(1): 3.
21. Lin L, Meng X, Liu P, Hong Y, Wu G, et al. (2009) Improved catalytic efficiency of endo-β-1,4-glucanase from *Bacillus subtilis* BME-15 by directed evolution. Appl Environ Microbiol 82: 671–679.
22. Andrews SR, Taylor EJ, Pell G, Vincent F, Ducros V, et al. (2004) The use of forced protein evolution to investigate and improve stability of family 10 xylanases. J Biol Chem 279(52): 54369–54379.
23. Navarro D, Couturier M, Silva GG, Berrin JG, Rouau X, et al. (2010) Automated assay for screening the enzymatic release of reducing sugars from micronized biomass. Microb Cell Fact 9: 58.
24. Couturier M, Roussel A, Rosengren A, Leone P, Stålbrand H, et al. (2013) Structural and biochemical analyses of glycoside hydrolase

families 5 and 26 beta-(1,4)-mannanases from *Podospora anserina* reveal differences upon manno-oligosaccharides catalysis. J Biol Chem 288(20): 14624–14635.

25. Duquesne S, Bordes F, Fudalej F, Nicaud JM, Marty A (2012) The yeast *Yarrowia lipolytica* as a generic tool for molecular evolution of enzymes. Methods Mol Biol 861: 301–312.
26. Emond S, Montanier C, Nicaud JM, Marty A, Monsan P, et al. (2010) New efficient recombinant expression system to engineer *Candida antartica* lipase B. Appl Environ Microbiol. 76(8): 2684–2687.
27. Liang C, Fioroni M, Rodríguez-Ropero F, Xue Y, Schwaneberg U, et al. (2011) Directed evolution of a thermophilic endoglucanase (Cel5A) into highly active Cel5A variants with an expanded temperature profile. J Biotechnol 154(1): 46–53.
28. Chen Y, Zhang B, Pei H, Lv J, Yang W, et al. (2012) Directed evolution of *Penicillium janczewskii* zalesk α-galactosidase toward enhanced activity and expression in *Pichia pastoris*. Appl Biochem Biotechnol 168(3): 638–650.
29. Boonvitthya N, Bozonnet S, Burapatana V, O'Donohue MJ, Chulalaksananukul W (2013) Comparison of the heterologous expression of *Trichoderma reesei*endoglucanase II and cellobiohydrolase II in the yeasts *Pichia pastoris* and *Yarrowia lipolytica*. Mol Biotechnol 54(2): 158–169.
30. Bordes F, Fudalej F, Dossat V, Nicaud JM, Marty A (2007) A new recombinant protein expression system for high-throughput screening in the yeast *Yarrowia lipolytica*. J Microbiol Meth 70(3): 493–502.
31. Nicaud JM, Madzak C, van den Broek P, Gysler C, Duboc P, et al. (2002) Protein expression and secretion in the yeast *Yarrowia lipolytica*. FEMS Yeast Res 2(3): 371–379.
32. Le Dall MT, Nicaud JM, Gaillardin C (1994) Multiple-copy integration in the yeast *Yarrowia lipolytica*. Curr Genet 26(1): 38–44.
33. Matsui I, Ishikawa K, Matsui E, Miyairi S, Fukui S, et al. (1991) Subsite structure of *Saccharomycopsis* alpha-amylase secreted from *Saccharomyces cerevisiae*. J Biochem 109: 566–569.

# CHAPTER 10

# APPLICATION OF A NOVEL ALKALI-TOLERANT THERMOSTABLE DYP-TYPE PEROXIDASE FROM SACCHAROMONOSPORA VIRIDIS DSM 43017 IN BIOBLEACHING OF EUCALYPTUS KRAFT PULP

**Wangning Yu[1], Weina Liu[1], Huoqing Huang[2] , Fei Zheng[1] , Xiaoyu Wang[1] , Yuying Wu[3] , Kangjia Li[1] , Xiangming Xie[1], Yi Jin[1]**

[1] College of Biological Sciences and Technology, Beijing Forestry University, Beijing, PR China

[2] Feed Research Institute Chinese Academy of Agricultural Sciences, Beijing, PR China

[3] College of Materials Science and Technology, Beijing Forestry University, Beijing, PR China

## ABSTRACT

*Saccharomonospora viridis* is a thermophilic actinomycete that may have biotechnological applications because of its dye decolorizing activity, though the enzymatic oxidative system responsible for this activity remains elusive. Bioinformatic analysis revealed a DyP-type peroxidase gene in the genome of *S. viridis* DSM 43017 with sequence similarity to peroxidase from dye-decolorizing microbes. This gene, *svidyp*, consists of 1,215 bp encoding a polypeptide of 404 amino acids. The gene encoding *Svi*DyP was cloned, heterologously expressed in *Escherichia coli*, and then purified. The recombinant protein could efficiently decolorize several triarylmethane dyes, anthraquinonic and azo dyes under neutral to alkaline conditions. The optimum pH and temperature for *Svi*DyP was pH 7.0 and 70°C, respectively. Compared with other DyP-type peroxidases, *Svi*DyP was more active at high temperatures, retaining>63% of its maximum activity at 50–80°C. It also showed broad pH adaptability (>35% activity at pH 4.0–9.0) and alkali-tolerance (>80% activity after incubation at pH 5–10 for 1 h at 37°C), and was highly thermostable (>60% activity after incubation at 70°C for 2 h at pH 7.0). *Svi*DyP had an accelerated action during the biobleaching of eucalyptus kraft pulp, resulting in a 21.8% reduction in kappa number and an increase of 2.98% (ISO) in brightness. These favorable properties make *Svi*DyP peroxidase a promising enzyme for use in the pulp and paper industries.

## INTRODUCTION

Heme peroxidases, containing prosthetic heme groups, have been divided into two large groups, animal and plant peroxidases [1]. The plant peroxidases are divided into classes I, II, and III, which include fungal (class II) and bacterial peroxidases (class I), according to primary structural homology [2]. Class I peroxidases contain mitochondrial yeast cytochrome c peroxidase, chloroplast and cytosol ascorbate peroxidases, and gene duplicated bacterial peroxidase [2]. Class II enzymes include secretory fungal peroxidases, such as lignin peroxidase [3], manganese peroxidase [4], and versatile peroxidase [5]. Secretory plant peroxidases such as horseradish peroxidase and barley grain peroxidase belong to class III peroxidases. The recently described DyP-type peroxidases (DyPs) (EC 1.11.1.19) superfamily comprises a novel class of heme peroxidases, and are found in both fungi and bacteria [6]–[14]. DyPs have been classified into four phylogenetically distinct subfamilies, with fungal enzymes belonging to subfamily D and bacterial enzymes constituting subfamilies A–C [15], [16]. These enzymes are very promising for use in biotechnological applications. Despite the large number of putative DyP sequences registered in PeroxiBase (http://peroxibase.toulouse.inra.

fr/), only a few studies have described the isolation and characterization of peroxidases, including four proteins from actinomycetes [17]-[21]. However, no DyP-type peroxidase genes have been identified in *Pseudonocardiaceae* to date.

Biobleaching is the bleaching of pulps using ligninolytic microorganisms or enzymes that reduce the amount of chemical bleach required to obtain a desirable brightness of pulps. Some enzymes from fungi and bacteria, such as extracellular xylanase and peroxidase from *Streptomyces* sp [22], [23], laccase and manganese peroxidase produced by *Trametes versicolor* [24], [25], have been studied for biobleaching of paper pulp and other industrial applications, as they degraded lignin or xylan and decolorized the pulp [26], [27]. However, no DyP-type peroxidase enzymes has been characterized and applied in the pulp bleaching field.

In the current work, a novel peroxidase, *Svi*DyP, related to a putative DyP-type peroxidase enzyme, was isolated, purified, and characterized from *Saccharomonospora viridis*, a pentachlorophenol-degrading thermophilic actinomycete [28]. In addition, the catalytic properties of the enzyme and its application as an enzymatic pretreatment for kraft pulp bleaching were investigated. *Svi*DyP was an alkali-tolerant thermostable dye-decolorizing peroxidase reported from the *Pseudonocardiaceae* family, which belongs to the genus of *Saccharomonospora*, and it had an accelerated action during the biobleaching of eucalyptus kraft pulp. These favorable properties made *Svi*DyP a good candidate for application in the pulp and paper industries.

# MATERIALS AND METHODS

## Microorganism, Culture Conditions, and Chemicals

*Saccharomonospora viridis* DSM 43017 was obtained from the China General Microbiological Culture Collection Center, Beijing (reference number CGMCC 4.1324). The strain was cultivated in a shaker flask at 45°C in STS medium (1.0% (w/v) soy peptone, 1.0% (w/v) glucose, 0.2% (w/v) yeast extract, 0.2% (w/v) NaCl, and 0.2% casein enzymatic hydrolysate, all from Biodee, Beijing, China), adjusted pH to 8.0 with NaOH prior to autoclaving.

*Escherichia coli* DH5α and *E. coli* BL21 (Tiangen, Beijing, China) (were cultivated in Luria Bertani (LB) medium at 37°C for gene cloning, sequencing, and expression. The pEASY-T3 vector (TransGen, Beijing, China) was used for plasmid gene cloning and sequencing. The plasmid pET-28a (+) (Takara Bio, Otsu, Japan) was used as an expression vector. Manganese peroxidase (MnP), azure B, brilliant green, reactive blue 19, reactive green 19, reactive yellow 2, reactive black 5, reactive red 120, malachite green and crystal violet were purchased from Sigma (St. Louis, MO, USA). All other reagents used were of analytical grade unless otherwise stated.

## Decolorization Assay for *S. viridis* on Solid Medium

Azure B, which was membrane-filtered with a 0.45-μm cellulose nitrate filter, was mixed with STS medium to give a final concentration of 0.1 g $l^{-1}$ and used to make agar plates. *S. viridis*was inoculated onto the plates with a sterile spreader. The plates were then incubated at 45°C and their surface appearance was observed daily. A ring of decoloration around the colonies on the blue medium is a qualitative signal of presence or absence of dye-decolorizing enzyme DyP-type peroxidase. The presence of bleaching around colonies on the blue medium indicates the production of the dye-decolorizing enzyme.

## Bioinformatic and Phylogenetic Analysis and Homology Modeling

The whole genome of *S. viridis* DSM 43017 is available from the GenBank database under the accession number NC_013159 [29]. By performing a sequence homology comparison between four well characterized DyP-type peroxidases from *Shewanella oneidensis* TyrA [13], *Bacteroides thetaiotaomicron* BtDyP [30], *Rhodococcus jostii* DyPB [31], and *E. coli* YcdB [32], one unannotated DyP-type peroxidase gene (Gene YP_003135694) was identified in the genome of *S. viridis* DSM 43017. The open reading frame, which we tentatively named *svidyp*, showed sequence similarity to peroxidase genes in other dye-decolorizing microbes. The DNA sequence of *svidyp* was analyzed using the software package DNAMAN 6.0 (Lynnon Corporation, Quebec, Canada). Nucleotide and the corresponding deduced amino acid sequence homology searches were carried out using the BLASTn and BLASTp programs, respectively, at the NCBI website (http://www.ncbi.nlm.nih.gov/BLAST/). Multiple sequence alignment was performed by ClustalW [33]. A phylogenetic tree containing the closest homologs of the DyP protein was constructed using MEGA 5.2 with the neighbor-joining algorithm [34]. The deduced tertiary structure of *Svi*DyP was derived from the amino acid sequence using SWISS-MODEL [35]–[37].

## Gene cloning, Heterologous Expression and Enzyme Purification

Genomic DNA from *S. viridis* was isolated following overnight growth in STS liquid medium. Cells were harvested by centrifugation (10,000×*g*, 25°C for 2 min). DNA extraction was then performed using the Bacterial Genomic DNA Extraction Kit V.3.0 (Takara) according to the manufacturer's instructions. Two primers were designed at the 5′ and 3′ ends of the *svidyp*gene with the following sequences: Svidyp-F: 5′-GGC<u>GAATTC</u>ATGAAGGGCCGGCGGTTC-3′and Svidyp-R: 5′-GCC<u>AAGCTT</u>TCAGGCTTCCAACAGCGG-3′, which contained the restriction sites *Eco*RI and *Hin*dIII (underlined), respectively, to enable directional cloning into the pET-28a (+) expression vector. Using these primers, the complete *svidyp* gene was PCR-amplified from *S. viridis* genomic DNA. PCR amplification was performed using *Taq* DNA polymerase, 10× PCR buffer ($Mg^{2+}$ Plus), and dNTP mixture (Takara) under the following conditions: 5 cycles of 94°C for 30 s, 67.3°C for 30 s, and 72°C for 1 min, followed by 30 cycles of 94°C for 30 s, 63.7°C for 30 s, and 72°C for 1 min.

The resulting PCR product was gel-purified, digested with *Eco*RI and *Hin*dIII, and cloned into the corresponding site of the pET28a (+) vector to construct the recombinant plasmid pET28a-svidyp. The plasmid was then transformed into chemically competent *E. coli* BL21 cells. Plasmid DNA from recombinant *E. coli* was isolated using a Plasmid Purification Kit (Takara). DNA concentration was estimated by the absorbance at 260 nm, and DNA integrity was verified by agarose gel electrophoresis. Transformants were identified by PCR analysis and enzymatic digestion, both as described above, and were then confirmed by DNA sequencing using the primers T7 and T7-terminor. Sequencing was carried out by Beijing AuGCT DNA-SYN Biotechnology Co., Beijing, China.

A positive transformant harboring pET28a-svidyp was isolated as a single colony for gene expression. The transformant was cultured overnight at 37°C in 10 ml LB medium containing kanamycin at 50 μg $ml^{-1}$. The culture was then inoculated into fresh LB medium (1:100 dilutions) containing kanamycin and was grown at 37°C to an $OD_{600}$ of approximately 0.6. Isopropyl-β-D-1-thiogalactopyranoside (IPTG) was then added to a final concentration of 1.0 mM for induction, and the culture was further cultivated at different temperature and time. Following induction with 1.0 mM IPTG at 25°C for 3 h, little activity was detected from the cell lysate, while incubation at 16°C for 12 h the activity in the lysate increased to 9.4 U $ml^{-1}$. Thus, to effectively induce enzyme expression, induction with 1.0 mM IPTG at 16°C overnight was selected for further production of the recombinant DyP peroxidise. No DyP peroxidase activity was detected from the uninduced transformant or from the transformant harboring the empty pET-28a (+) vector.

Cultures of induced recombinant *E. coli* (200 ml) were centrifuged (10,000×*g*, 4°C for 10 min) and the cell pellets were resuspended in 10 ml sodium phosphate buffer (50 mM, pH 8.0). The cell suspension was then lysed by sonication (pause on: 10 s; pause off: 10 s; power: 130 W; time: 30 min) using a cell ultra sonicator (Sonics* VCX130; Sonics & Materials, CT, USA). Following sonication, the unbroken cells and large debris were removed by centrifugation (12,000×*g*, 4°C for 20 min) and the supernatant was collected. Because the enzyme was expressed as a His-tagged protein, the supernatant was loaded onto a Ni-affinity column (Ni–NTA Agarose; Qiagen, Valencia, CA, USA) and eluted with a linear gradient of imidazole (20–250 mM, 25 ml) in sodium phosphate buffer (20 mM phosphate, 300 mM NaCl, pH 8.0), at a flow rate of 1 ml $min^{-1}$. Fractions displaying DyP-peroxidase activity were combined and applied to a Sephadex-G75 (10×200 mm) column pre-equilibrated with 0.1 M phosphate buffer (pH 7.0) and eluted with the same buffer at a flow rate of 0.5 ml $min^{-1}$. The purified enzyme with purification yields of 40.3% (Table 1) was stored at −20°C until further characterization. The protein concentration was determined using a Bradford assay [38] using bovine serum albumin as a standard. The purity of *Svi*DyP was verified by sodium dodecyl sulfate polyacrylamide gel electrophoresis (SDS-PAGE) analysis on a 12% gel as described previously [39] and by Western blotting analysis using an anti-His6 tag mouse monoclonal antibody [40].

**Table 1:** Purification of recombinant *Svi*DyP

https://doi.org/10.1371/journal.pone.0110319.t001

| Step | Activity (U) | Protein (mg) | Specific activity (U mg $^{-1}$) | Yield (%) | Purification fold |
|---|---|---|---|---|---|
| Lysate supernatant | 940.0 | 1175 | 0.8 | 100.0 | 1.0 |
| Ni-NTA agarose | 437.9 | 39.1 | 11.2 | 46.6 | 14.0 |
| Sephadex G-75 | 379.1 | 21.3 | 17.8 | 40.3 | 22.3 |

doi:10.1371/journal.pone.0110319.t001

# Enzyme Assays

Brilliant green, a representative triarylmethane dye, was used as the substrate. The substrate solution consisted of 125 μM brilliant green in 50 mM phosphate buffer (pH 7.0). One hundred microliters of the purified *Svi*DyP enzyme solution (0.01 mg $ml^{-1}$) was mixed with the substrate solution. All reactions were started by addition of $H_2O_2$ (0.1 mM for all tests). The total volume of the enzyme reaction mixture was then adjusted to 3 ml. The enzyme activity was calculated by measuring the variation in absorbance at 623 nm using the molar extinction coefficient of brilliant green (max. wavelength: 623 nm; $\varepsilon_{623}$=18,000$M^{-1}cm^{-1}$). One unit of enzyme activity was defined as the amount of enzyme that decolorized 1 μmol of brilliant green at 60°C in 1 min [41]. The same method was used to study the substrate range of the enzyme using reactive blue 19 (max. wavelength: 590 nm; $\varepsilon_{590}$=10,000 $M^{-1}$ $cm^{-1}$), reactive green 19 (max. wavelength: 630 nm; $\varepsilon_{630}$=22,000 $M^{-1}$ $cm^{-1}$), reactive yellow 2 (max. wavelength: 390 nm; $\varepsilon_{390}$=8,000 $M^{-1}$ $cm^{-1}$), reactive black 5 (max. wavelength: 598 nm; $\varepsilon_{598}$=30,000 $M^{-1}$ $cm^{-1}$), reactive red 120 (max. wavelength: 535 nm; $\varepsilon_{535}$=28,000 $M^{-1}$ $cm^{-1}$), malachite green (max. wavelength: 617 nm; $\varepsilon_{617}$=8,400 $M^{-1}$ $cm^{-1}$), crystal violet (max. wavelength: 583 nm; $\varepsilon_{583}$=21,900 $M^{-1}$ $cm^{-1}$), azure B (max. wavelength: 651 nm; $\varepsilon_{630}$=22,000 $M^{-1}$ $cm^{-1}$), 2, 6 - Dimethoxyphenol (max. wavelength: 469 nm; $\varepsilon_{469}$=2,750 $M^{-1}$ $cm^{-1}$) and veratryl alcohol (max. wavelength: 310 nm; $\varepsilon_{310}$=9,300 $M^{-1}$ $cm^{-1}$) as substrates under their corresponding maximum absorption wavelength. Control treatments without $H_2O_2$ and/ or without enzyme were performed. MnP activity was measured at 25°C by the formation of $Mn^{+3}$-tartrate complex ($\varepsilon_{238}$=6,500 $M^{-1}$ $cm^{-1}$) during the oxidation of 0.5 mM $MnSO_4$ in 100 mM tartrate bufferat pH 4.5 with 0.1 mM $H_2O_2$ [8]. One unit of MnP activity corresponds to the amount of enzyme which oxidizes 1 μmol $Mn^{2+}$ to $Mn^{3+}$ per minute at pH 4.5 and 25°C (in the presence of 0.1 mM $H_2O_2$).

## Enzyme Characterization

Aliquots of the recombinant *Svi*DyP sample (1 mg $ml^{-1}$) were put under various conditions to test its stability. The optimal pH for enzyme activity was determined by measuring the reaction rate in buffers covering a pH range of 3.0–12.0. The buffers used were 50 mM citric acid-$Na_2HPO_4$ (pH 3.0–8.0), 50 mM Tris-HCl (pH 8.0–9.0), and 50 mM glycine-NaOH (pH 9.0–12.0). The pH stability was measured by adding enzyme to buffers ranging from pH 2.0–10.0 at 37°C for 1 h without the substrate solutions, and then, its residual activity was analyzed as described above. The highest activity of the enzyme, treated as above, was defined as 100%. The optimal temperature for enzyme activity was examined over the temperature range of 30–90°C. To estimate the thermal stability, the enzymes were incubated at 50°C, 60°C, 70°C, or 80°C for 0.5, 1, 2, 3, 4, and 5 h without substrate, and the residual enzyme activity was measured with the method as described above. To determine the effect of different metal ions and chemical reagents on the enzyme activity, the enzyme sample (1 mg $ml^{-1}$) was incubated for 30 min at 37°C and pH 7.0 in 0.05 M Britton–Robinson buffer, supplemented with the following additives respectively (final concentration, 5 mM): SDS, β-mercaptoethanol, ethylenediaminetetraacetic acid (EDTA), $ZnCl_2$, $BaCl_2$, $MnCl_2$, $MgCl_2$, $AlCl_3$, $CoCl_2$, $CaCl_2$, LiCl, $FeCl_2$, and $FeCl_3$. The residual activity was then assayed using the method as described above. And a system without any additives was used as a control.

The $K_m$, and $V_{max}$ values for the purified *Svi*DyP protein were determined by several substrates including reactive blue 19, reactive yellow 2 and brilliant green. Lineweaver-Burk plots were generated from the initial rates obtained at varying substrate concentrations with a constant concentration of the second substrate [42]. Each experiment was repeated three times and each experiment included three replicates.

### Physico-Chemical Characterization of Kraft Pulps and Biobleaching with *Svi*DyP

An unbleached eucalyptus kraft pulp was prepared by cooking the dried wood material (provided by JinHai Pulp & Paper Limited Company,

Hainan, China) with sulfate process under the following conditions: total alkali, 16.0% (w/v), sulfureted degree 25% (w/v), liquid ratio 1:6, and a cooking temperature of 160°C for 2 h. The optimization of enzyme dose and reaction time for biobleaching was carried out by treating pulp with varying doses of DyP-type peroxidase, ranging from 0–6.0 U $g^{-1}$ dry pulp, for various time intervals up to 6 h. Pulp properties were investigated at regular intervals. Enzyme-treated and untreated pulp samples at 10% pulp consistency were incubated at 65°C with *Svi*DyP (5 milligram protein per gram dried pulp, equivalently enzyme dosage as 3 U $g^{-1}$ which was in terms of enzyme activity towards brilliant green) in 50 mM citric acid-$Na_2HPO_4$ buffer (pH 7.0) for 2 h, whereas the biobleaching by MnP with 0.5 mM $MnSO_4$ was performed at 25°C in pH 4.5 for 2 h as the positive control. 0.1 mM $H_2O_2$ was added in all the assays. During the bleaching treatment, kneading the pulp mixture every 15 min was needed in order to make sure all ingredients can mix well. After the enzyme bleaching, pulps were thoroughly washed with deionized water to remove soluble reducing sugars and any free soluble residual lignin, and then made into pieces of paper by the sheet forming machine. Pulp sheet preparation was conducted as described in the paper [43]. Degree of brightness of pulp according to ISO (International Standards Organization) was measured on an YQ-Z-48A (Qingtong, Hangzhou, China) brightness color tester. Pulp kappa number was determined by using TAPPI useful method T236 (Technical Association of the Pulp and Paper Industry, Atlanta, Ga.). All reported data are averages of experiments performed in at least six replicates.

## Scanning electron Microscopy

Samples of pulp fibers were processed for scanning electron microscopy (SEM). Paper pulp samples were formed into hand sheets using a Buchner funnel. Small pieces of fiber were vacuum dried in the presence of $P_2O_5$, placed on the stubs and mounted with silver tape, and then sputter coated with gold using fine coat. The samples were examined at 5.0 kV under a scanning electron microscope (Hitachi S-4300, Ibaraki, Japan) at various magnifications.

# RESULTS

## Decolorization Assay for *S. viridis* on Solid Medium

Some bleaching laps were observed on STS medium mixed with 0.1 g $l^{-1}$ azure B around colonies of *S. viridis* (Fig. 1), which were produced after incubating at 45°C for 3 days, indicating the presence of a dye-decolorizing enzyme. This indicated one or more extracellular dye-decolorizing enzymes produced in the cultivating process of *S. viridis*.

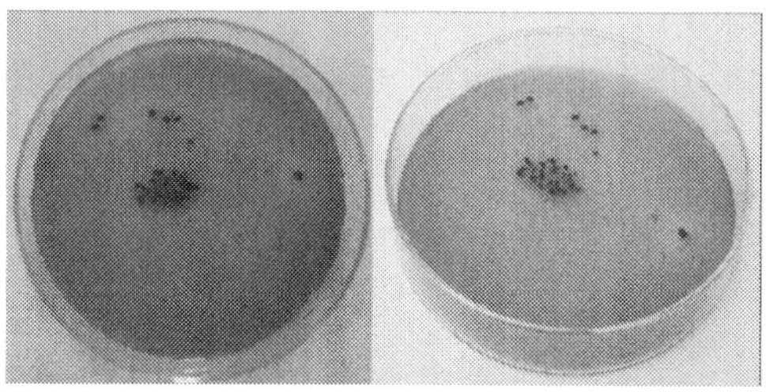

**Figure 1:** Degradation circles produced by *Saccharomonospora viridis* **on azure B screening plates.**

https://doi.org/10.1371/journal.pone.0110319.g001

## *Svi*DyP is a Member of the DyP-type Family Enzymes

To isolate the gene encoding the DyP-type peroxidase from *S. viridis* DSM 43017, two specific primers, Svidyp-F and Svidyp-R, were designed. PCR amplification of the DyP-peroxidase gene (GenBank under accession number KF444221) of *S. viridis* generated a DNA fragment of 1215 bp, which encoded a polypeptide of 404 amino acids with a predicted molecular mass of 44.5 kDa and a deduced isoelectric point of 5.17. Sequence analysis showed that overall G + C content of the *svidyp* gene was 70.1%.

The deduced amino acid sequence of *Svi*DyP showed 36% amino acids identity with *Bacillus subtilis* BsDyP [15] and *R. jostii* DyPB [31], 33% *E. coli* YcdB [32], 27% to *S. oneidensis* TyrA [13] and 21% to *B. thetaiotaomicron* BtDyP [30]. Twenty typical and hotspot DyPs proteins, such as *Thermobifida fusca* TfuDyP [21], *S. oneidensis* TyrA [13], *B. subtilis* BsDyP [15], *and R. jostii* DyPB [31], were used to phylogenetic analysis. Phylogenetic trees and multiple sequence alignments showed that *Svi*DyP was most closely related to *Mycobacterium* sp. *My*spDyP, *Thermobifida fusca Tfu*DyP, *Rhodococcus jostii* DyPA, *Escherichia coli* EfeB/YcdB, *Bacillus subtilis Bs*DyP [15] and the putative uncharacterized protein Sco3963 from *Streptomyces coelicolor* A3 (2), all of which belong to subfamily A (Fig 2). Therefore the enzyme *Svi*DyP was grouped into subfamily A, and typical motifs, such as conserved GXXDG motif and some conserved amino acid residues (D223, N227, R328, H313, E342, D371), were found in DyP-type peroxidases (Fig 3). The model structure of *Svi*DyP showed two domains consisting of a four-stranded antiparallel β-sheet and peripheral α-helices which is typical in the DyP-type family.

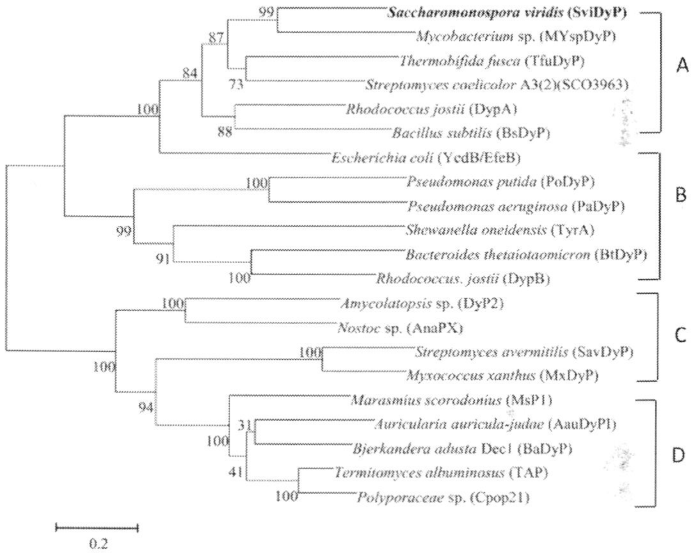


**Figure 2:** Phylogenetic tree analysis.

Phylogenetic tree showing the relationship between *Svi*DyP and other isolated DyPs belonging to subfamilies A–D: MYspDyP from *Mycobacterium* sp. (GI: 126434170), *Tfu*DyP from *Thermobifida fusca* (GI: 71917209), Sco3963 from *Streptomyces coelicolor*A3(2) (GI: 541881521), DypA from *Rhodococcus jostii* (GI: 499917154), *Bs*DyP from *Bacillus subtilis* (GI: 732344), EfeB/YcdB from *Escherichia coli* (GI: 2506638), *Po*DyP from *Pseudomonas putida* (GI: 501229266), *Pa*DyP from *Pseudomonas aeruginosa* (GI: 94829180), TyrA from *Shewanella oneidensis* (GI: 119390160), *Bt*DyP from *Bacteroides* *thetaiotaomicron* (GI: 109158102), DypB from *R. jostii* (GI: 330689635), DyP2 from *Amycolatopsis* sp. (GI: 409973917), AnaPX from *Nostoc* sp. (GI: 75704119), *Sav*DyP from *Streptomyces avermitilis* (GI: 29604188), *Mx*DyP from *Myxococcus xanthus* (GI: 108465542), Msp1 from *Marasmius scorodonius* (GI: 261266601), *Aau*DyPI from *Auricularia auricula-judae* (GI: 433286646), TAP from *Termitomyces albuminosus* (GI: 20386144), and Cpop21 from *Polyporaceae* sp. (GI: 2160705). Confidence was evaluated with 1000 bootstrap replicates (values are indicated at branch points).

https://doi.org/10.1371/journal.pone.0110319.g002

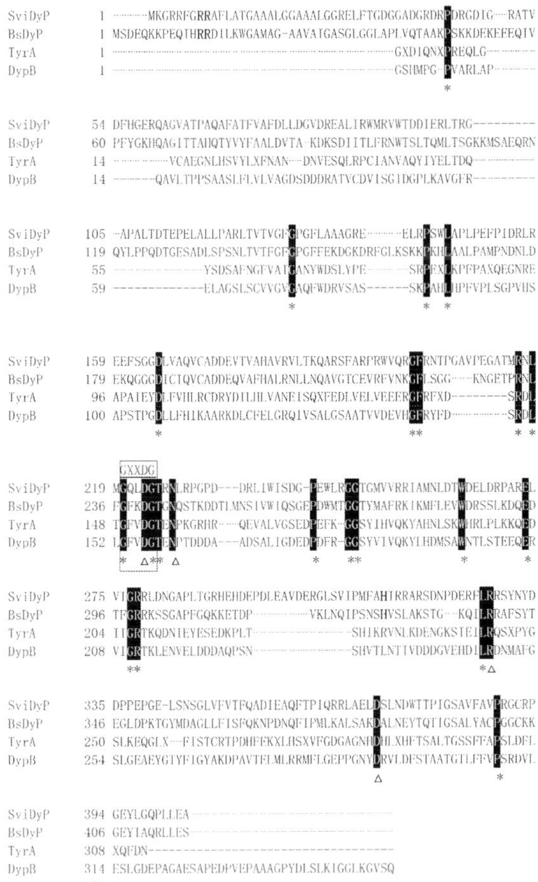

**Figure 3:** Multiple sequence alignment of the SviDyP amino acid sequence with other members of DyP-type peroxidases family.

The sequence alignments were performed using ClustalW (Thompson et al. 1994). Perfect match residues are displayed in white on black shade. The unique GXXDG motif is boxed. The conserved residues of the active site are indicated by asterisks. Residues indicated by triangles are residuals important for coordinating the heme at the active site.

https://doi.org/10.1371/journal.pone.0110319.g003

## Heterologous Expression and Purification of *Svi*DyP

The svidyp gene was successfully cloned into the pET28a vector and expressed in *E. coli*BL21, forming the recombinant clone *E. coli* BL21/ pET28a-svidyp, with the expression of His-tagged protein. The purified *Svi*DyP protein, as verified by SDS-PAGE analysis and Western Blot analysis (Fig. 4a) by anti-6xHis antibody, was about 45 kDa, consistent with the theoretical value of 44.5 kDa.

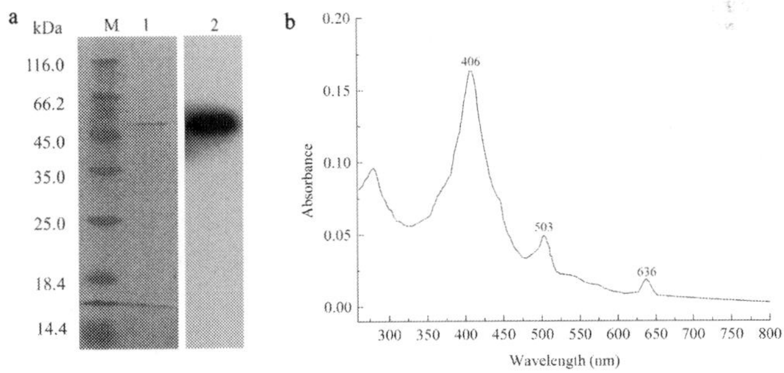


**Figure 4:** SDS–PAGE and western blotting analysis and spectral characteristics of purified *Svi*DyP.

**a**. SDS–PAGE and western blotting analysis. Lanes: M, molecular mass standard; (1) SDS–PAGE analysis of purified *Svi*DyP. (2) western blot analysis of the purified *Svi*DyP recognized by an anti-His6 tag mouse monoclonal antibody. **b**. Spectral characteristics of purified *Svi*DyP, as determined by hemoprotein assay. Purified *Svi*DyP (1 mg $ml^{-1}$) in 25 mM phosphate buffer (pH 7.0) was scanned at 300–800 nm to identify the Soret band.

https://doi.org/10.1371/journal.pone.0110319.g004

## Biochemical and Spectroscopic Characterization of *Svi*DyP

The UV-visible spectrum of *Svi*DyP at resting state showed the main Soret band at 406 nm and two charge transfer bands at 503 nm and 636 nm (Fig. 4b), as well as a Reinheitszahl ratio ($A_{406}/A_{280}$) of 1.7. The molar absorption coefficient of *Svi*DyP, $\varepsilon_{406,}$ was 97,000 $M^{-1}$ $cm^{-1}$, similar to values found for other peroxidases and DyPs [8], [44].

The recombinant enzyme expressed and purified from *E. coli* BL21 was then biochemically characterized. The controls of without $H_2O_2$ and/or without enzyme were without significant statistical difference. Substrate specificity of the enzyme was determined by using a subset of well-known peroxidase substrates and anthraquinone and azo dyes (Table 2). The enzyme displayed maximal activity towards the anthraquinone dye reactive blue 19 (1.29 U $mg^{-1}$), the azo dyes reactive green 19, reactive yellow 2, reactive black 5, and reactive red 120 (1.32, 4.86, 0.96 and 0.69 U $mg^{-1}$, respectively), and the triarylmethane dyes brilliant green, malachite green, and crystal violet (12.24, 8.4 and 4.11 U $mg^{-1}$, respectively). *Svi*DyP was able to act on veratryl alcohol (0.03 U $mg^{-1}$), but not very efficiently (Table 2), and also showed activity towards azure B (1.62 U $mg^{-1}$), a methylated thiazine dye that is a good substrate for lignin peroxidase [45]. Additionally, *Svi*DyP showed a modest activity towards 2, 6 - dimethoxyphenol (0.06 U $mg^{-1}$) which is a typical methoxylated phenol substrate for plant peroxidases. From the above, it is true that *Svi*DyP displays (DyP-type) peroxidase activity.

**Table 2:** Enzymatic activity of purified *Svi*DyP towards various dyes[a]

https://doi.org/10.1371/journal.pone.0110319.t002

| Substrate | $\lambda_{max}$ (nm)/ε at $\lambda_{max}$[b] ($M^{-1}$ $cm^{-1}$) | pH | Temp (°C) | Initial concn (μM) | Specificity activity (U $mg^{-1}$)[c] | Relative activity (%)[d] |
|---|---|---|---|---|---|---|
| Reactive Blue 19[e] | 590/10000 | 7.0 | 70 | 50 | 1.29 | 10.5 |
| Reactive Green 19[f] | 630/22000 | 7.0 | 60 | 35 | 1.32 | 10.8 |
| Reactive Yellow 2[f] | 390/8000 | 7.0 | 60 | 125 | 4.86 | 39.7 |
| Reactive Black 5[f] | 598/30000 | 7.0 | 60 | 12.5 | 0.96 | 7.8 |
| Reactive Red 120[f] | 535/28000 | 7.0 | 60 | 12.5 | 0.69 | 5.6 |
| Brilliant Green[g] | 623/18000 | 7.0 | 70 | 12.5 | 12.24 | 100 |
| Malachite Green[g] | 617/8400 | 7.0 | 70 | 32 | 8.4 | 68.6 |
| Crystal Violet[g] | 583/21900 | 7.0 | 70 | 50 | 4.11 | 33.6 |
| Azure B | 651/8700 | 6.0 | 70 | 32 | 1.62 | 13.2 |
| 2,6-Dimethoxyphenol | 469/2750 | 4.5 | 60 | 100 | 0.06 | 0.5 |
| Veratryl alcohol | 310/9300 | 4.5 | 60 | 100 | 0.03 | 0.2 |

[a]The total volume of the enzyme reaction mixture was 3 ml.
[b]Molar extinction coefficient at maximum absorption wavelength (λmax) of each dye.
[c]Activity was calculated as specific activity in U $mg^{-1}$ (1 U = 1 μmol/min).
[d]Relative activity was defined as activity towards brilliant green.
[e]AQ, anthraquinone.
[f]AZ, azo.
[g]TM, triarylmethane.
doi:10.1371/journal.pone.0110319.t002

To study the behavior of the enzyme under various pH and temperature conditions, brilliant green was used as the substrate. The optimal pH of *Svi*DyP at 60°C was 7.0. It had broad pH adaptability (>35% activity at pH 4.0–9.0) and retained over 62% of the peak activity at pH 6.0–8.0 (Fig. 5a). *Svi*DyP was highly alkali-tolerant, retaining at least 80% of its initial activity after incubation at 37°C for 1 h in buffers ranging from pH 5.0–10.0 (Fig. 5b). The optimal temperature for *Svi*DyP activity was 70°C, at pH 7.0 (Fig. 5c). *Svi*DyP was highly thermophilic, retaining at least 63% of its maximum activity at 50–80°C, and approximately 95% at 80°C. The thermostability of the enzyme was investigated for a period of 5 h at temperatures ranging from 50–80°C (Fig. 5d). It was observed that *Svi*DyP maintained approximately 60% of maximal activity at 60°C and 70°C for at least 3 h and 2 h, respectively.

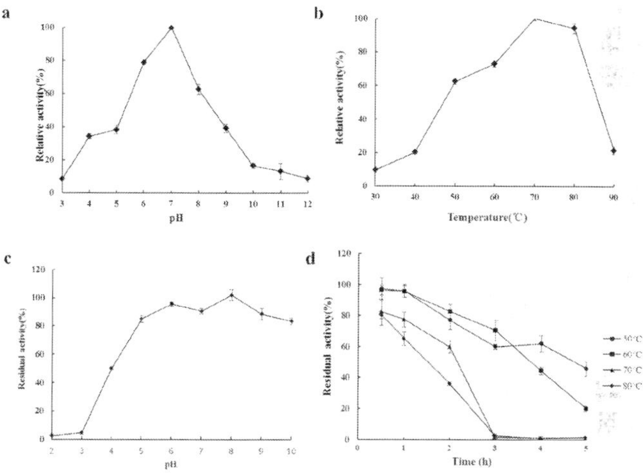


**Figure 5:** Characterization of *Svi*DyP.

a. pH optimization. The enzyme was assayed at different pH, as described in the "Materials and Methods"section. b. Temperature optimization. The enzyme samples were incubated in pH 7.0 at 30, 40, 50, 60, 70, 80, and 90°C for 30 min and their residual activities were assayed, as described in the "Materials and Methods"section. c. pH stability. Enzyme samples (1 mg ml$^{-1}$) were incubated for 30 min at different pH levels at 37°C, and then their residual activities were assayed at the optimum pH,as described in the "Materials and Methods" section. d. Temperature stability. The enzyme samples were incubated in pH 7.0

at 50, 60, 70, or 80°C, and then their residual activities were assayed at different times.

https://doi.org/10.1371/journal.pone.0110319.g005

The effect of different metal ions and metal chelating agents on the enzymatic activity was also assayed, using brilliant green as the substrate (Table 3). The activity of control without any additives was defined as 100%. The activity of the enzyme was not affected (residual activity,>80%) by the presence of 5 mM final concentration of $Ba^{2+}$, $Mn^{2+}$, $Mg^{2+}$, $Fe^{2+}$, or EDTA. An obvious reduction in catalytic activity was found in solutions containing 5 mM $Fe^{3+}$ or SDS. $Al^{3+}$ and $Co^{2+}$ caused 27.6% and 16.9% activation of *Svi*DyP, respectively. The addition of 5 mM $Cu^{2+}$, $Zn^{2+}$, $Ca^{2+}$, or $Li^{+}$ had similar but less activating effects on the activity of *Svi*DyP, probably through promoting an oxidation reaction. β-mercaptoethanol highly boosted the enzymatic activity by almost a factor of 8.5.

**Table 3:** Effect of various agents on *Svi*DyP activity

https://doi.org/10.1371/journal.pone.0110319.t003

| Reagent | Residual activity (%) 5 mM |
|---|---|
| None | 100 |
| $Zn^{2+}$ | 102.3±2.0 |
| $Ba^{2+}$ | 99.6±1.7 |
| $Mn^{2+}$ | 92.3±2.1 |
| $Mg^{2+}$ | 96.4±3.1 |
| $Al^{3+}$ | 127.6±3.0 |
| $Co^{2+}$ | 116.9±2.2 |
| $Ca^{2+}$ | 84.7±4.1 |
| $Li^{+}$ | 87.0±3.3 |
| $Fe^{2+}$ | 96.1±4.1 |
| $Fe^{3+}$ | 35.9±3.9 |
| EDTA | 97.2±2.0 |
| β-mercaptoethanol | 847.9±1.9 |
| SDS | UD |

UD, undetectable activity.
EDTA, ethylenediaminetetraacetic acid.
SDS, sodium dodecyl sulfate.
doi:10.1371/journal.pone.0110319.t003

The kinetic analysis of purified *Svi*DyP corroborated the Michaelis-Menten behavior of the enzyme, with a $V_{max}$ of 38.46±0.98 μmol $l^{-1}$ $min^{-1}$

and a $K_m$ of 58.87±4.7 μmol $l^{-1}$ for brilliant green, a $V_{max}$ of 3.7±0.41 μmol $l^{-1}$ $min^{-1}$ and a $K_m$ of 187.2±16.6 μmol $l^{-1}$ for reactive yellow 2, and a $V_{max}$ of 2.2±0.26 μmol $l^{-1}$ $min^{-1}$ and a $K_m$ of 210.2±16.5 μmol $l^{-1}$ for reactive blue 19. Such kinetic behavior strongly suggested that the cloned enzyme might be a DyP-type peroxidase.

## Enzyme effects on Pulp Bleaching

Enzyme effects on pulp bleaching were reflected by measuring the physicochemical properties of paper pulp biobleached by enzymic preparations. The unbleached/raw eucalyptus kraft pulp had a kappa number of 11.0 and a brightness of 33.6% (ISO) [43]. With the increase of enzyme dosage and reaction time, the pulp brightness was increased and kappa number was reduced gradually, and then they keep invariant. When the enzyme dosage was 3 U $g^{-1}$ (, b) and the reaction time was 2 h, the pulp brightness reached the maximum, and pulp kappa number reached the lowest value. Therefore, the choice of 3 U $g^{-1}$ as an optimal enzyme dosage and 2 h as an optimal reaction time were the best condition of *Svi*DyP biobleaching kraft pulp. The experiments of pulp bleaching by recombinant protein *Svi*DyP showed that enzyme treatment to improve the brightness of the pulp 2.98%, pulp kappa number reduced 21.8% compared with control (Table 4), indicating that the enzyme treatment can significantly improve the brightness of the pulp, but reduce kappa value will help subsequent bleaching. Application of MnP in the pulp bleaching experiment showed that MnP reduced the specific pulp kappa number by 31.8%, and increased the whiteness of pulp by 5.90% at pH 4.5, 25°C (Table 4), whereas it reduced the specific pulp kappa number by 24.8% and increased the whiteness of pulp by 3.26% at pH 7, 65°C.

**Table 4:** Physicochemical properties of enzyme-treated eucalyptus kraft pulp

https://doi.org/10.1371/journal.pone.0110319.t004

| Enzyme (Treatment) | Samples | Brightness (% ISO) | Brightness increase (%) | Kappa number | Kappa decrease (%) |
|---|---|---|---|---|---|
| *Svi*DyP | Control[a] | 33.6±0.1 | - | 11.0±0.2 | - |
| (65°C pH 7.0) | Test | 34.6±0.4 | 2.98 | 8.6±0.1 | 21.8 |
| MnP | Control | 33.7±0.4 | - | 10.9±0.3 | - |
| (65°C pH 7.0) | Test | 34.8±0.3 | 3.26 | 8.2±0.1 | 24.8 |
| MnP | Control | 33.9±0.2 | - | 11.0±0.4 | - |
| (25°C pH 4.5) | Test | 35.9±0.2 | 5.90 | 7.5±0.2 | 31.8 |

[a]The control samples and the test samples were incubated in the selected buffer at the set temperature with and without corresponding enzyme.
doi:10.1371/journal.pone.0110319.t004

To study the effect of the enzyme on fiber surface morphology in *Svi*DyP-bleached pulp, eucalyptus kraft pulp fiber morphology and surface changes were observed by SEM (Fig. 6). Without any enzymatic bleaching, eucalyptus kraft pulp fibers were stiff, and the surface structure was relatively smooth with no damage or cracks (Fig. 6a). The surface morphology of fibers treated with MnP is shown in Fig. 6b and the fibers treated with DyP-type peroxidase from *S. viridis* are shown in Fig. 6c and 6d. The pulp fiber surface treated with DyP-type peroxidase from *S. viridis* appeared rough and wrinkled and showed signs of slight fissures as well as that treated with MnP. There were also many holes in the surface fibers, and fibers began to increase in flexibility (Fig. 6b, 6c and 6d). These representative images show the obvious cracks, grooves and spalling in the surface of the fibers, with some fibers having holes with an aperture diameter of 7 μm, which increases fiber flexibility.

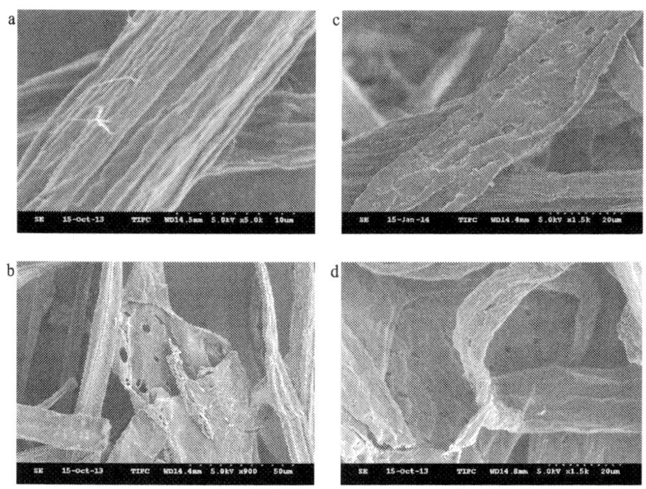

**Figure 6:** Morphology of bleached pulp fiber surface.

Scanning electron micrograph of unbleached (control) eucalyptus kraft pulp (a). Scanning electron micrograph of MnP treated eucalyptus kraft pulp (b). Scanning electron micrograph of *Svi*DyP treated eucalyptus kraft pulp (c, d).

https://doi.org/10.1371/journal.pone.0110319.g006

# DISCUSSION

The pentachlorophenol-degrading thermophilic actinomycete *S. viridis* and its extracellular enzymes offer great potential for different biotechnological applications. Both a crude extracellular xylanase [46] and a purified xylanase [47], apparently in the absence of cellulase or endoglucanase activity, have been described in *S. viridis*, and may represent a potential new enzyme source for use in pulp bleaching preparations. In the current study, we reported the cloning, expression, purification, characterization, and pulp biobleaching application of an alkali-tolerant thermostable peroxidase *Svi*DyP from *S. viridis* DSM 43017.

Bioinformatic analysis revealed one unannotated DyP-type peroxidase gene in the genome of *S. viridis* DSM 43017 that had sequence similarity with other DyPs at the primary structure level showing the characteristic conserved residues in the heme-binding site and GXXDG motif [12]. Based on the phylogenetic analysis, *Svi*DyP belongs to subfamily A of DyPs. Compared with other DyPs from subfamily A, the purified recombinant *Svi*DyP, appeared remarkably broad range substrate specificity of which could degrade various dyes, such as triarylmethane dyes, anthraquinonic and azo dyes, at different levels, much better than some enzymes grouped into subfamily A, for example, YcdB [48], TfuDyP [21]. The enzyme could efficiently decolorize triarylmethane dyes to other triarylmethane dye-degrading microorganisms [49], [50], and was more active towards crystal violet [51], [52] and malachite green [51], [53], [54]. *Svi*DyP also showed the ability to degrade anthraquinonic and azo dyes under neutral to alkaline conditions. Hence, such a multifunctional peroxidase with wide substrate specificity was likely capable of decolorizing other, currently unidentified, xenobiotics, Which suggested that *Svi*DyP might be applied directly to solve various environmental and industrial problems, such as the disposal of dye wastewater and pulp biobleaching rather than the DyPs [8], [9] which usually worked under acid condition.

To further explore the feasibility of using *Svi*DyP in biotechnological applications, enzyme stability was evaluated at different pH levels and temperatures. More luckily, *Svi*DyP was very stable over a pH range of 5–10, and retained more than 80% of the maximum residual activity. It also displayed high thermal stability, maintaining 63% of residual activity after 4 h of incubation at 50°C, and more than 60% of residual activity after 2 h of incubation at 60°C. *Svi*DyP showed greater stability than DyP from *Anabaena* sp. [16], which had an activity loss of 90% following 3 h incubation at 50°C. Stability studies showed that *Svi*DyP exhibited a high degree of stability under alkaline pH and high temperatures. This ability to work efficiently under such high pH environments was a characteristic that distinguishes DyPs from most peroxidases [55]. These findings enriched DyP-type peroxidases family's information databases and laid the foundation for the study of structure, function and reaction mechanisms of *Svi*DyP.

Concerning the influence factors of using SviDyP in the further bleaching industrial application, the activation/inactivation effect of the different metal ions and chemical reagents on the enzyme activity was studied. Interestingly, addition of EDTA did not inhibit the enzyme, which suggested that there are no essential ions in the reaction mixture for the enzyme activity [7], [8]. $Al^{3+}$ had similar activating effect between *Svi*DyP and *Pseudomonas aeruginosa* DyPPa [7], probably through promoting the oxidation. SDS resulted in the complete activity loss of *Svi*DyP and DyPPa [7], which demonstrated that SDS could be an important DyP inhibitor. β-mercaptoethanol highly boosted the enzymatic activity presumably by counteracting the oxidation effects of the S–S linkage between cysteine residues [56], [57].

The applicability of *Svi*DyP in biobleaching of kraft pulp, a process of high biotechnological interest, was examined by prebleaching eucalyptus kraft pulp with the purified *Svi*DyP enzyme. Under 65°C, pH 7.0, *Svi*DyP enzyme had positive effects on the pulp bleaching process as well as MnP, while the biobleaching effects of MnP at 25°C, pH 4.5, were better than *Svi*DyP at 65°C, pH 7.0. These data suggest a slightly temperature and pH adjustment were needed in the usage of *Svi*DyP in bleaching pulp. Furthermore, the SEM studies revealed that the existence of local fracture phenomena and the large area of the trench structure containing holes would result in the fiber structure becoming loose, and the internal structure of lignin being fully exposed. This may help the bleaching agent permeate better, and perfects delignification, which will happen later. Thereby, this breakdown in structure means that much less bleaching agent was needed, and less poisoned organochlorine was produced (data not shown).

In addition, the prepared *Svi*DyP peroxidase had hardly some cellulase activity (data not shown), thereby it prevented reduction of fiber strength and viscosity of thick liquid. The enzyme did show significant heat and alkali resistance, and could directly and effectively attack lignin. Therefore, *Svi*DyP is applicable to the process of pre-bleaching paper pulp for industrial production, and shows excellent bleaching effects. With its wide substrate specificities and thermophilic capabilities, *Svi*DyP peroxidase is a promising enzyme with considerable biotechnological and commercial potential, especially in the pulp and paper industry.

## AUTHOR CONTRIBUTIONS

Conceived and designed the experiments: XX WY YJ. Performed the experiments: WY WL. Analyzed the data: WY FZ. Contributed reagents/ materials/analysis tools: KL XW YW HH. Wrote the paper: WY WL. Revised the first draft of the manuscript: HH. Read, corrected, and approved the final manuscript: XX YJ.

## REFERENCES


1. Dunford HB (1999) Heme peroxidases. Wiley-VCH.
2. Welinder KG (1992) Superfamily of plant, fungal and bacterial peroxidases. Curr Opin Struc Biol 2: 388–393.
3. Tien M, Kirk TK (1988) Lignin peroxidase of *Phanerochaete chrysosporium*. Method Enzymol 161: 238–249.
4. Wariishi H, Valli K, Gold MH (1992) Manganese (II) oxidation by manganese peroxidase from the basidiomycete *Phanerochaete chrysosporium*. Kinetic mechanism and role of chelators. J Biol Chem 267: 23688–23695.
5. Pogni R, Baratto MC, Giansanti S, Teutloff C, Verdin J, et al. (2005) Tryptophan-based radical in the catalytic mechanism of versatile peroxidase from *Bjerkandera adusta*. Biochemistry 44: 4267–4274.
6. Colpa DI, Fraaije MW, van Bloois E (2014) DyP-type peroxidases: a promising and versatile class of enzymes. J Ind Microbiol Biotechnol 41: 1–7.
7. Li J, Liu C, Li B, Yuan H, Yang J, et al. (2012) Identification and molecular characterization of a novel DyP-type peroxidase from

*Pseudomonas aeruginosa*PKE117. Appl Biochem Biotech 166: 774–785.

8. Salvachúa D, Prieto A, Martínez ÁT, Martínez MJ (2013) Characterization of a novel dye-decolorizing peroxidase (DyP)-Type Enzyme from *Irpex lacteus* and its application in enzymatic hydrolysis of wheat straw. Appl Environ Microbiol 79: 4316–4324.
9. Sato T, Hara S, Matsui T, Sazaki G, Saijo S, et al. (2004) A unique dye-decolorizing peroxidase, DyP, from *Thanatephorus cucumeris* Dec 1: heterologous expression, crystallization and preliminary X-ray analysis. Acta Crystallogr D 60: 149–152.
10. Coughlin MF, Kinkle BK, Bishop PL (2003) High performance degradation of azo dye acid orange 7 and sulfanilic acid in laboratory scale reactor after seeding with cultured bacterial strains. Water Res 37: 2757–2763.
11. Shimokawa T, Shoda M, Sugano Y (2009) Purification and characterization of two DyP isozymes from *Thanatephorus cucumeris* Dec 1 specifically expressed in an air-membrane surface bioreactor. J Biosci Bioeng 107: 113–115.
12. Sugano Y (2009) DyP-type peroxidases comprise a novel heme peroxidase family. Cell Mol Life Sci 66: 1387–1403.
13. Zubieta C, Joseph R, Sri Krishna S, McMullan D, Kapoor M, et al. (2007) Identification and structural characterization of heme binding in a novel dye-decolorizing peroxidase, TyrA. Proteins: Structure, Function, and Bioinformatics 69: 234–243.
14. Liers C, Pecyna MJ, Kellner H, Worrich A, Zorn H, et al. (2013) Substrate oxidation by dye-decolorizing peroxidases (DyPs) from wood-and litter-degrading agaricomycetes compared to other fungal and plant heme-peroxidases. Appl Microbiol Biotechnol 97: 5839–5849.
15. Santos A, Mendes S, Brissos V, Martins LO (2014) New dye-decolorizing peroxidases from *Bacillus subtilis* and *Pseudomonas putida* MET94: towards biotechnological applications. Appl Microbiol Biotechnol 98: 2053–2065.
16. Ogola HJO, Kamiike T, Hashimoto N, Ashida H, Ishikawa T, et al. (2009) Molecular characterization of a novel peroxidase from

the cyanobacterium *Anabaena* sp. strain PCC 7120. Appl Environ Microbiol 75: 7509–7518.

17. Ahmad M, Roberts JN, Hardiman EM, Singh R, Eltis LD, et al. (2011) Identification of DypB from *Rhodococcus jostii* RHA1 as a lignin peroxidase. Biochemistry 50: 5096–5107.
18. Fodil D, Badis A, Jaouadi B, Zaraî N, Ferradji FZ, et al. (2011) Purification and characterization of two extracellular peroxidases from *Streptomyces* sp. strain AM2, a decolorizing actinomycetes responsible for the biodegradation of natural humic acids. Int Biodeter Biodegr 65: 470–478.
19. Fodil D, Jaouadi B, Badis A, Nadia ZJ, Ferradji FZ, et al. (2012) A thermostable humic acid peroxidase from *Streptomyces* sp. strain AH4: Purification and biochemical characterization. Bioresource Technol 111: 383–390.
20. Roberts JN (2011) Identification and characterization of DyP peroxidases from *Rhodococcus jostii* RHA1: University of British Columbia.
21. van Bloois E, Pazmiño DET, Winter RT, Fraaije MW (2010) A robust and extracellular heme-containing peroxidase from *Thermobifida fusca* as prototype of a bacterial peroxidase superfamily. Appl Microbiol Biotechnol 86: 1419–1430.
22. Antonopoulos V, Hernandez M, Arias M, Mavrakos E, Ball A (2001) The use of extracellular enzymes from *Streptomyces albus* ATCC 3005 for the bleaching of eucalyptus kraft pulp. Appl Microbiol Biotechnol 57: 92–97.
23. Georis J, Giannotta F, De Buyl E, Granier Bt, Frère J-M (2000) Purification and properties of three endo-β-1, 4-xylanases produced by *Streptomyces* sp. strain S38 which differ in their ability to enhance the bleaching of kraft pulps. Enzyme Microb Tech 26: 178–186.
24. Wesenberg D, Kyriakides I, Agathos SN (2003) White-rot fungi and their enzymes for the treatment of industrial dye effluents. Biotechnol Adv 22: 161–187.
25. Reid ID, Paice MG (1998) Effects of manganese peroxidase on residual lignin of softwood kraft pulp. Appl Environ Microbiol 64: 2273–2274.

26. Beg QK, Bhushan B, Kapoor M, Hoondal G (2000) Enhanced production of a thermostable xylanase from *Streptomyces* sp. QG-11-3 and its application in biobleaching of eucalyptus kraft pulp. Enzyme Microb Techn 27: 459–466.

27. Harazono K, Kondo R, Sakai K (1996) Bleaching of hardwood kraft pulp with manganese peroxidase from *Phanerochaete sordida* YK-624 without addition of MnSO (inf4). Appl Environ Microbiol 62: 913–917.

28. Webb MD, Ewbank G, Perkins J, McCarthy AJ (2001) Metabolism of pentachlorophenol by *Saccharomonospora viridis* strains isolated from mushroom compost. Soil Biol Biochem 33: 1903–1914.

29. Pati A, Sikorski J, Nolan M, Lapidus A, Copeland A, et al. (2009) Complete genome sequence of *Saccharomonospora viridis* type strain (P101$^{T}$). Stand Genomic Sci 1: 141–149.

30. Zubieta C, Krishna S, Kapoor M, Kozbial P, McMullan D, et al. (2007) Crystal structures of two novel dye-decolorizing peroxidases reveal a β-barrel fold with a conserved heme-binding motif. Proteins: Structure, Function, and Bioinformatics 69: 223–233.

31. Roberts JN, Singh R, Grigg JC, Murphy ME, Bugg TD, et al. (2011) Characterization of dye-decolorizing peroxidases from *Rhodococcus jostii* RHA1. Biochemistry 50: 5108–5119.

32. Liu X, Du Q, Wang Z, Zhu D, Huang Y, et al. (2011) Crystal structure and biochemical features of EfeB/YcdB from *Escherichia coli* O157 ASP235 plays divergent roles in different enzyme-catalyzed processes. J Biol Chem 286: 14922–14931.

33. Thompson JD, Higgins DG, Gibson TJ (1994) CLUSTAL W: improving the sensitivity of progressive multiple sequence alignment through sequence weighting, position-specific gap penalties and weight matrix choice. Nucl Acids Res 22: 4673–4680.

34. Kumar S, Nei M, Dudley J, Tamura K (2008) MEGA: a biologist-centric software for evolutionary analysis of DNA and protein sequences. Brief Bioinform 9: 299–306.

35. Arnold K, Bordoli L, Kopp J, Schwede T (2006) The SWISS-MODEL workspace: a web-based environment for protein structure homology modelling. Bioinformatics 22: 195–201.

36. Kiefer F, Arnold K, Künzli M, Bordoli L, Schwede T (2009) The

SWISS-MODEL repository and associated resources. Nucl Acids Res 37: D387–D392.

37. Schwede T, Kopp J, Guex N, Peitsch MC (2003) SWISS-MODEL: an automated protein homology-modeling server. Nucl Acids Res 31: 3381–3385.
38. Bradford MM (1976) A rapid and sensitive method for the quantitation of microgram quantities of protein utilizing the principle of protein-dye binding. Anal Biochem 72: 248–254.
39. Laemmli UK (1970) Cleavage of structural proteins during the assembly of the head of bacteriophage T4. Nature 227: 680–685.
40. Cao P, Tang XM, Guan ZB, Diao ZY, Zhang SQ (2005) Production and characterization of a bacterial single-chain antibody fragment specific to B-cell-activating factor of the TNF family. Protein Expr Purif 43: 157–164.
41. Sugano Y, Nakano R, Sasaki K, Shoda M (2000) Efficient heterologous expression in *Aspergillus oryzae* of a unique dye-decolorizing peroxidase, DyP, of *Geotrichum candidum* Dec 1. Appl Environ Microbiol 66: 1754–1758.
42. Lineweaver H, Burk D (1934) The determination of enzyme dissociation constants. J Am Chem Soc 56: 658–666.
43. Kondo R, Harazono K, Sakai K (1994) Bleaching of hardwood kraft pulp with manganese peroxidase secreted from *Phanerochaete sordida* YK-624. Appl Environ Microbiol 60: 4359–4363.
44. Kim SJ, Shoda M (1999) Purification and characterization of a novel peroxidase from *Geotrichum candidum* Dec 1 involved in decolorization of dyes. Appl Environ Microbiol 65: 1029–1035.
45. Archibald FS (1992) A new assay for lignin-type peroxidases employing the dye azure B. Appl Environ Microbiol. 58: 3110–3116.
46. Roberts JC, McCarthy AJ, Flynn NJ, Broda P (1990) Modification of paper properties by the pretreatment of pulp with *Saccharomonospora viridis* xylanase. Enzyme Microb Tech 12: 210–213.
47. Wang Z, Jin Y, Wu H, Tian Z, Wu Y, et al. (2012) A novel, alkali-tolerant thermostable xylanase from *Saccharomonospora viridis*: direct gene cloning, expression and enzyme characterization. World J Microbiol Biotechnol 28: 2741–2748.

48. Sturm A, Schierhorn A, Lindenstrauss U, Lilie H, Brüser T (2006) YcdB from *Escherichia coli* reveals a novel class of Tat-dependently translocated hemoproteins. J Biol Chem 281: 13972–13978.
49. An S-Y, Min S-K, Cha I-H, Choi Y-L, Cho Y-S, et al. (2002) Decolorization of triphenylmethane and azo dyes by *Citrobacter* sp. Biotechnol Lett 24: 1037–1040.
50. Ren S, Guo J, Zeng G, Sun G (2006) Decolorization of triphenylmethane, azo, and anthraquinone dyes by a newly isolated *Aeromonas hydrophila* strain. Appl Microbiol Biotechnol 72: 1316–1321.
51. Sani RK, Banerjee UC (1999) Decolorization of triphenylmethane dyes and textile and dye-stuff effluent by *Kurthia* sp. Enzyme and Microbial Technology 24: 433–437.
52. Yatome C, Yamada S, Ogawa T, Matsui M (1993) Degradation of crystal violet by *Nocardia corallina*. Appl Microbiol Biotechnol 38: 565–569.
53. Daneshvar N, Ayazloo M, Khataee A, Pourhassan M (2007) Biological decolorization of dye solution containing Malachite Green by microalgae *Cosmarium* sp. Bioresource Technol 98: 1176–1182.
54. Jones JJ, Falkinham JO (2003) Decolorization of malachite green and crystal violet by waterborne pathogenic mycobacteria. Antimicrob Agents Chemother 47: 2323–2326.
55. Sugano Y, Muramatsu R, Ichiyanagi A, Sato T, Shoda M (2007) DyP, a unique dye-decolorizing peroxidase, represents a novel heme peroxidase family ASP171 replaces the distal histidine of classical peroxidases J Biol Chem. 282: 36652–36658.
56. Heck JX, Flores SH, Hertz PF, Ayub MA (2006) Statistical optimization of thermo-tolerant xylanase activity from Amazon isolated *Bacillus circulans* on solid-state cultivation. Bioresource Technol 97: 1902–1906.
57. Du Y, Shi P, Huang H, Zhang X, Luo H, et al. (2013) Characterization of three novel thermophilic xylanases from *Humicola insolens* Y1 with application potentials in the brewing industry. Bioresource Technol 130: 161–167.

# CHAPTER 11

# REDUCTION IN ACUTE ECOTOXICITY OF PAPER MILL EFFLUENT BY SEQUENTIAL APPLICATION OF XYLANASE AND LACCASE

**Saurabh Sudha Dhiman[1,2], Gaurav Garg[3,4], Jitender Sharma[3], Vipin C. Kalia[5], Yun Chan Kang[1], JungKul Lee[1,2]**

[1] Department of Chemical Engineering, Konkuk University, Gwangjin–Gu, Seoul, Republic of Korea

[2] Institute of SK–KU Biomaterials, Konkuk University, Gwangjin–Gu, Seoul, Republic of Korea

[3] Department of Biotechnology, Kurukshetra University, Kurukshetra, Haryana, India

[4]Department of Biotechnology, Maharishi Markandeshwar University, Mullana-Ambala, Haryana, India

[5] Microbial Biotechnology and Genomics, CSIR-Institute of Genomics and Integrative Biology, Delhi University Campus, Delhi, India

## ABSTRACT

In order to reduce the ecotoxicity of paper mill, four different enzymatic pretreatment strategies were investigated in comparison to conventional chemical based processes. In strategy I, xylanase-aided pretreatment of pulp was carried out, and in strategy II, xylanase and laccase-mediator systems were used sequentially. Moreover, to compare the efficiency of *Bacillus stearothermophilus* xylanase and *Ceriporiopsis subvermispora* laccase in the reduction of ecotoxicity and pollution, parallel strategies (III and IV) were implemented using commercial enzymes. Conventional $C_DE_{OP}D_1D_2$ ($C_{D,}$ $Cl_2$ with $ClO_2$; $E_{OP,}$ $H_2O_2$ extraction; $D_1$ and $D_{2,}$ $ClO_2$) and $X/XLC_DE_{OP}D_1D_2$ (X, xylanase; L, laccase) sequences were employed with non-enzymatic and enzymatic strategies, respectively. Acute toxicity was determined by the extent of inhibition of bioluminescence of *Vibrio fischeri* with different dilutions of the effluent. Two-fold increase was observed in $EC_{50}$ values for strategy I compared to the control process. On the other hand, sequential application of commercial enzymes resulted in higher acute toxicity compared to lab enzymes. In comparison to the control process, strategy II was the most efficient and successfully reduced 60.1 and 25.8% of biological oxygen demand (BOD) and color of effluents, respectively. We report for the first time the comparative analysis of the ecotoxicity of industrial effluents.

## INTRODUCTION

In the last two decades, use of enzymes, especially hemicellulases, has revolutionized the pulp and paper industry and provided a glimpse of hope that application of enzymes at various levels can reduce the industrial pollution and effluent's toxicity. However, the current scenario continues to be challenging because of the high pollution load released by the pulp and paper industries, which are still using chlorine-based bleaching sequences [1]. The conventional bleaching methodology is chemical dependent and energy intensive; therefore, a novel enzymatic treatment with safe level of discharge needs to be developed [2]. Hydrolytic enzymes along with a laccase mediator system (LMS) have been more beneficial in reducing the pollution load of industries compared to other strategies [3]. An N-hydroxy-based synthetic mediator was predominantly used for this purpose [4]. However, toxicity and cost are two of the major hurdles, which hamper the industrial applications of these synthetic mediators. Therefore, the application of natural mediators in LMS is one of the alternatives to overcome these disadvantages, even though their application may cause grafting onto the pulp, an increase in kappa number, and a reduction in brightness of the pulp. Since chemical-intensive conventional strategies and enzymatic processes utilizing synthetic mediators release high levels of toxic compounds into water bodies, the entire processing of the effluents should be characterized to analyze their ecotoxicity and other hazardous properties [5].

It has been observed that pretreatment with xylanase alone cannot reduce the pollution load of pulp and paper industry significantly. Hence, it is believed that a cocktail of two or more enzymes could reduce the release of hazardous materials to safer levels [4]. Although enzymes are effective at the pretreatment level, effluents from the entire process should be analyzed to study the enzymatic after-effects. Few reports deal with the characterization of effluents from enzyme-aided bleaching processes, but without evaluating the interaction of technical parameters at the pretreatment level [2]. Therefore, in the present investigation, the individual and cumulative effects of physical parameters on the efficiency of both xylanase and laccase were optimized and were analyzed using a statistical model.

This is the first ever attempt where ecotoxicity of mixed effluents from the entire bleaching process was characterized using Microtox 81.9% basic toxicity assay method along with the evaluation of the reduction in pollution load in terms of biological oxygen demand (BOD) and color. A schematic study was done with four different strategies by supplementing the conventional bleaching sequence with xylanase from *Bacillus stearothermophilus* SDX and Pulpzyme VLBL (Novozyme, Denmark). In sequential strategies, *Ceriporiopsis subvermispora*laccase and the commercial laccase TM L603P (UK) were used with a natural mediator (syringaldehyde, SA) for the extraction of cellulosic fibers from agro-residual material for paper processing.

# MATERIALS AND METHODS

## Microbial Strains and Their Maintenance

The bacterial culture for xylanase production was isolated from a compost sample and it was identified by the Institute of Microbial Technology (IMTECH), Chandigarh, India, as *B. stearothermophilus* SDX and strain has been deposited at IMTECH and the accession number is 8508 [6]. A lignin-degrading fungal culture of *C. subvermispora* L-14807 SS-3 was used for laccase production [2].

## Enzyme Production, Extraction and Assay

Xylanase production was studied through two-step statistical modeling [7] under solid-state fermentation (SSF) conditions. An Erlenmeyer flask (250 mL) containing 5 g of wheat bran as sole carbon source was moistened with 25 mL of modified Horikoshi medium (pH 8.0) [8]. Sterilized substrate, cooled to room temperature, was inoculated with 15% (v/w) inoculum (18 h culture, ~$3.6\times10^6$ cells/mL) and incubated at 37°C in a humidified chamber (relative humidity 60–65%) for 96 h. The flasks were gently tapped intermittently to mix the contents. Colonized solid culture (1 g) was extracted twice with 10 mL of phosphate buffer (10 mM, pH 8.0) through a wet muslin cloth. Xylanase activity was determined through modified Bailey's method [9] by using 3,5-dinitrosalicylic acid (DNS) reagent [10]. All the parameters related to laccase production were optimized through the conventional "one variable at a time" method under submerged fermentation (SmF) conditions. The crude extract of *C. subvermispora* culture was partially purified using ammonium sulfate fractionation (50–70% cut) and was concentrated with a 30-kDa cutoff membrane filter (Amicon). The observed molecular weight of the laccase was approximately 50 kDa. To obtain the concentrated partially purified enzyme, a membrane with a 30-kDa cutoff was used. The laccase activity was determined according to a previously reported method [11]. Crude extracts of both the strains were assayed for other enzymatic activities viz. cellulase and amylase. Similar to xylanase from *B. stearothermophilus* SDX and laccase from *C. subvermispora*, commercial xylanase and laccase were also characterized.

Estimation of the protein content in all the enzyme samples was determined by Bradford's method [12]. All the experiments were carried out independently in triplicates and the results are presented as the mean of the three values.

## Screening of Natural Mediators

Different phenolic compounds such as acetosyringone (AS), coniferyl alcohol (CA), *p*-coumaric acid (PCA), 2,6-dimethoxyphenol (DMP),

2,6-dimethylphenol (DP), guaiacol, 4-hydroxybenzoic acid (HBA), 4-hydroxybenzaldehyde (4-HA), and syringaldehyde (SA) were screened to determine their potential as natural mediators that facilitate decolorization. The potential of natural phenols (50 μM each) was determined using the decolorization assay of 25 μM Azure-B [13]. The catalytic efficiency of natural phenols was compared with that of the synthetic mediator 1-hydroxybenzatriazole (HBT) as a control.

## Enzyme, Chemicals and Software

Commercial xylanase (Pulpzyme VLBL) and laccase (TM L603P) were kindly provided by Novozyme (Denmark) and Treforest Industrial Estate (Pontypridd, CF375 UD, Wales, United Kingdom), respectively. Agro-residual pulp composed of (w/w) wheat straw (*Triticum aestivum*) (78.8%), sarkanda (*Saccharum spontaneum*) (10.6%), and candy (*Eragrostis* sp.) (10.6%) cooked at 165–175°C for 30 min at a pressure of 7.0–7.5 kg $m^{-3}$, was obtained from an agro-based industry (Trident Paper, Barnala, Punjab) in India. The permanganate number, brightness (%ISO), yellowness, and viscosity (poise) of the unbleached pulp were 7.11±0.96, 42.5±5.6, 15.8±2.6, and 5.18±0.96, respectively. Experimental design was generated using the statistical software "Design-Expert 6.0" developed by Stat-Ease Inc. (Minneapolis, MN, USA) to study the individual and cumulative effect of independent variables on enzyme production and biobleaching.

## Parametric Optimization for Pretreatment of Pulp

Pretreatment of pulp with xylanase and laccase was carried out separately. Different levels of all the independent variables designed through central composite design of response surface methodology (CCD-RSM) are given. For xylanase-aided pretreatment, 10% (w/v) pulp consistence was used and statistical modeling was applied to four independent variables viz. pH, temperature ($T_X$), enzyme dose ($ED_X$), and retention time ($RT_{X,}$ Table 1).

**Table 1:** Experimental design (coded variables) and results of CCD-RSM for xylanase-aided pretreatment of pulp

https://doi.org/10.1371/journal.pone.0102581.t001

| pH | $T_X$[a] | $ED_X$[b] | $RT_X$[c] | $PN_X$[d] Actual | $PN_X$[d] Predicted | $B_X$[e] Actual | $B_X$[e] Predicted | $Y_X$[f] Actual | $Y_X$[f] Predicted |
|---|---|---|---|---|---|---|---|---|---|
| 0 | 0 | 0 | -α | 6.92±1.72 | 6.82±1.13 | 43.1±5.7 | 41.5±4.2 | 14.2±3.1 | 14.3±1.5 |
| +1 | +1 | -1 | +1 | 7.12±1.44 | 7.31±1.32 | 48.5±5.7 | 44.7±4.5 | 14.8±2.4 | 15.0±1.6 |
| -α | 0 | 0 | 0 | 6.74±1.31 | 6.85±1.15 | 44.7±6.4 | 42.8±4.3 | 14.2±2.1 | 14.1±1.5 |
| +1 | +1 | +1 | -1 | 7.21±1.32 | 7.09±1.24 | 39.5±4.5 | 40.5±4.2 | 14.5±2.9 | 14.6±1.5 |
| 0 | 0 | 0 | 0 | 6.58±1.35 | 6.50±0.92 | 45.8±2.9 | 45.8±4.1 | 13.5±1.2 | 13.5±1.3 |
| -1 | +1 | +1 | -1 | 6.74±1.23 | 6.69±0.82 | 42.1±5.4 | 42.7±4.3 | 13.5±1.9 | 14.1±1.4 |
| 0 | 0 | 0 | 0 | 6.58±1.24 | 6.50±0.91 | 45.8±5.7 | 45.8±4.5 | 13.5±3.1 | 13.5±1.3 |
| 0 | +α | 0 | 0 | 6.94±1.22 | 6.85±0.92 | 43.4±5.6 | 43.9±4.1 | 14.1±2.5 | 14.6±1.4 |
| 0 | 0 | +α | 0 | 6.66±1.35 | 6.87±0.92 | 44.5±1.5 | 41.4±4.4 | 13.8±3.4 | 14.3±1.5 |
| 0 | 0 | -α | 0 | 7.78±1.33 | 7.33±1.34 | 37.2±5.4 | 38.4±3.1 | 13.4±1.9 | 14.7±1.4 |
| -1 | +1 | -1 | -1 | 6.73±1.24 | 6.78±1.25 | 42.1±6.4 | 41.0±4.2 | 13.5±2.3 | 14.0±1.4 |
| +α | 0 | 0 | 0 | 7.82±1.42 | 7.35±1.24 | 38.5±9.4 | 39.3±3.3 | 16.9±3.1 | 15.7±1.6 |
| -1 | -1 | +1 | -1 | 6.97±1.85 | 6.74±0.92 | 42.2±4.6 | 45.5±4.4 | 14.2±2.4 | 14.4±1.5 |
| +1 | -1 | -1 | +1 | 7.24±1.33 | 7.26±1.12 | 39.6±2.9 | 40.3±3.1 | 14.9±1.2 | 14.9±1.5 |
| +1 | +1 | +1 | +1 | 6.98±1.34 | 6.93±1.11 | 42.5±4.7 | 43.3±4.2 | 14.1±2.6 | 14.4±1.6 |
| 0 | 0 | 0 | 0 | 6.53±1.22 | 6.50±0.81 | 45.8±6.8 | 45.8±4.3 | 13.5±2.8 | 13.8±1.3 |
| -1 | -1 | +1 | +1 | 7.05±1.25 | 6.88±0.82 | 42.2±2.5 | 43.0±4.4 | 14.0±3.2 | 13.9±1.4 |
| 0 | 0 | 0 | 0 | 6.53±1.34 | 6.50±0.81 | 45.8±4.3 | 45.8±4.1 | 13.5±2.1 | 13.9±1.3 |
| +1 | +1 | -1 | -1 | 7.12±1.22 | 7.28±1.12 | 39.5±6.8 | 40.1±3.2 | 14.3±3.2 | 14.7±1.5 |
| -1 | -1 | -1 | +1 | 7.04±1.34 | 7.16±1.12 | 42.0±5.7 | 41.4±4.3 | 14.1±3.5 | 14.2±1.5 |
| +1 | -1 | +1 | +1 | 6.96±1.15 | 6.88±0.93 | 39.9±6.8 | 40.6±4.5 | 14.6±3.1 | 14.4±1.5 |
| 0 | 0 | 0 | 0 | 6.58±1.42 | 6.50±0.81 | 45.8±3.8 | 45.8±4.2 | 13.5±3.8 | 13.9±1.3 |
| 0 | -α | 0 | 0 | 6.99±1.34 | 6.85±0.81 | 43.7±5.4 | 42.3±4.4 | 14.2±1.3 | 14.1±1.4 |
| -1 | -1 | -1 | -1 | 6.82±1.32 | 6.83±0.82 | 41.5±4.8 | 42.1±4.3 | 14.7±2.5 | 14.1±1.4 |
| +1 | -1 | -1 | -1 | 7.01±1.55 | 7.16±1.22 | 38.7±3.4 | 38.1±3.1 | 14.6±2.2 | 14.7±1.5 |
| -1 | +1 | +1 | +1 | 6.94±1.51 | 6.78±0.91 | 42.5±4.6 | 42.7±4.5 | 14.1±2.3 | 13.8±1.4 |
| 0 | 0 | 0 | 0 | 6.54±1.63 | 6.50±0.82 | 45.8±4.5 | 45.8±4.1 | 13.5±2.1 | 13.9±1.4 |
| -1 | +1 | -1 | +1 | 7.16±1.28 | 7.06±1.12 | 41.2±5.8 | 42.8±4.2 | 14.9±3.2 | 14.3±1.5 |
| +1 | -1 | +1 | -1 | 6.95±1.62 | 6.99±1.02 | 40.5±4.6 | 40.3±4.2 | 14.2±3.2 | 14.8±1.5 |
| 0 | 0 | 0 | +α | 7.02±1.14 | 6.98±0.81 | 42.8±5.3 | 43.5±4.2 | 14.1±2.4 | 13.8±1.4 |

a - Temperature (°C); b - Enzyme Dose (U/mg); c - Retention time (min); d - Permanganate Number; e - Brightness (%ISO) of pulp after pretreatment; f - Yellowness (b*) of pulp after pretreatment.

doi:10.1371/journal.pone.0102581.t001

After extensive washing with distilled water, xylanase-treated pulp was subjected to LMS. A broad range of reaction parameters (viz. enzyme dose [$ED_L$], mediator concentrations [$MC_L$], and retention times [$RT_L$]) were optimized through CCD-RSM for the LMS (Table 2). The pulp was again washed thoroughly with distilled water before determining its strength. Similar to xylanase pretreatment, regression equation and responses were determined for the pretreatment of agro-residual pulp with the LMS.

**Table 2:** Experimental design (coded variables) and results of CCD-RSM for laccase-aided pretreatment of pulp

https://doi.org/10.1371/journal.pone.0102581.t002

| Run | $RT_L$[a] | $ED_L$[b] | $MC$[c] | $PN_L$[d] Actual | $PN_L$[d] Predicted | $B_L$[e] Actual | $B_L$[e] Predicted | $Y_L$[f] Actual | $Y_L$[f] Predicted |
|---|---|---|---|---|---|---|---|---|---|
| 1 | 0 | 0 | 0 | 5.69±0.82 | 5.23±0.53 | 47.6±4.1 | 47.8±4.5 | 12.8±1.2 | 12.7±1.1 |
| 2 | +1 | -1 | -1 | 5.68±0.82 | 5.43±0.52 | 47.5±4.2 | 49.2±4.5 | 11.9±1.4 | 10.6±1.0 |
| 3 | -1 | +1 | +1 | 5.41±0.61 | 4.76±0.42 | 48.2±4.2 | 47.0±4.5 | 11.5±2.1 | 11.5±1.3 |
| 4 | 0 | -α | 0 | 5.85±0.73 | 5.23±0.51 | 46.8±4.2 | 48.4±3.4 | 13.2±1.6 | 12.9±1.4 |
| 5 | 0 | 0 | 0 | 5.69±0.81 | 6.10±0.42 | 47.6±4.4 | 46.0±4.5 | 12.8±1.3 | 12.5±1.2 |
| 6 | -1 | +1 | -1 | 5.89±0.72 | 5.46±0.51 | 47.8±4.1 | 46.7±4.5 | 13.2±3.1 | 12.3±1.2 |
| 7 | -1 | -1 | -1 | 5.92±0.74 | 4.88±0.45 | 48.1±3.8 | 47.2±4.5 | 13.8±1.4 | 11.4±1.0 |
| 8 | 0 | +α | 0 | 5.46±0.72 | 5.46±0.43 | 48.6±3.9 | 46.4±4.6 | 11.3±1.3 | 11.7±1.7 |
| 9 | +1 | +1 | -1 | 5.41±0.81 | 5.95±0.44 | 48.2±3.6 | 46.6±5.4 | 11.6±1.2 | 12.9±1.2 |
| 10 | +1 | -1 | +1 | 6.52±0.83 | 4.86±0.61 | 47.6±4.1 | 47.4±5.5 | 11.6±1.4 | 11.5±1.2 |
| 11 | +1 | +1 | +1 | 5.23±0.83 | 5.66±0.61 | 48.9±3.6 | 46.4±4.5 | 10.9±1.4 | 12.3±1.3 |
| 12 | 0 | 0 | 0 | 5.69±0.72 | 5.23±0.52 | 47.6±3.9 | 48.4±4.5 | 12.8±1.5 | 12.9±1.8 |
| 13 | +α | 0 | 0 | 5.36±0.72 | 5.22±0.52 | 48.5±4.1 | 47.0±4.5 | 11.5±1.4 | 11.5±1.3 |
| 14 | 0 | 0 | 0 | 5.67±0.72 | 6.03±0.52 | 47.6±3.6 | 47.2±4.5 | 13.8±1.5 | 12.6±1.3 |
| 15 | -1 | -1 | +1 | 6.03±0.71 | 5.23±0.51 | 46.7±4.1 | 47.2±4.5 | 12.9±1.1 | 12.2±1.6 |
| 16 | 0 | 0 | 0 | 5.69±0.72 | 5.23±0.33 | 47.6±3.8 | 48.4±4.5 | 12.8±1.2 | 12.9±1.2 |
| 17 | 0 | 0 | 0 | 5.69±0.71 | 5.23±0.41 | 47.6±4.1 | 48.4±4.5 | 12.8±1.3 | 12.9±1.2 |
| 18 | -α | 0 | 0 | 6.24±0.83 | 5.23±0.52 | 45.9±3.7 | 47.1±4.4 | 13.7±1.2 | 13.1±1.1 |
| 19 | 0 | 0 | +α | 5.37±0.71 | 4.76±0.43 | 48.9±3.4 | 46.9±5.4 | 10.9±1.1 | 11.2±1.3 |
| 20 | 0 | 0 | -α | 5.15±0.72 | 5.29±0.62 | 45.4±3.2 | 47.6±5.4 | 13.3±1.2 | 14.0±1.3 |

a - Retention Time (min); b - Enzyme Dose (U/mg); c - Mediator Concentration (%); d - Permanganate Number; e - Brightness (%ISO) of pulp after C. subvermispora laccase mediated pretreatment; f - Yellowness (b*) of pulp after C. subvermispora laccase mediated pretreatment

doi:10.1371/journal.pone.0102581.t002

## Residual Laccase Activity

Laccase activity was determined under controlled conditions using 50 g oven-dried (OD) pulp and SA as mediator. Residual laccase activity was determined in the presence and absence of pulp and mediators and was expressed as percentage of initial activity, which was measured at the beginning (zero time) of the incubation time. Experiments were performed using 20 U of laccase and 1.5 mM of mediator.

## Scheme ($X/XLC_DE_{OP}D_1D_2$) for Biobleaching

The most extensively used bleaching sequence $C_DE_{OP}D_1D_2$ ($C_D$, chlorine and chlorine dioxide; $E_{OP}$, extraction with oxygen and hydrogen peroxide; $D_1$ and $D_2$, chlorine dioxide) was used as the control. In enzyme-aided strategies, only xylanase-assisted stage (X) and pretreatment with both xylanase and laccase along with natural mediator step (XL) were included prior to the conventional sequence.

Dry plastic bags filled with 50 g of extensively washed OD pulp were subjected to different combinations of reaction parameters as designed using CCD-RSM. During chlorination stage, the consistency of the pulp was maintained at 3% for effective penetration of the fiber by the reaction liquor. Experiments were designed with different doses of $Cl_2$ at the $C_D$ stage, and a kappa factor of 0.25 was used for calculating the $Cl_2$ dose. The $E_{OP}$ stage was carried out in a stainless steel vessel at 70±1°C after adding the required amounts of sodium hydroxide (25 kg/T) and hydrogen peroxide ($H_2O_2$; 5 kg/T). The oxygen pressure was maintained at 5.0 and 2.5 kg/cm$^2$ for the initial 30 min and final 90 min, respectively. Final pH in the reaction vessel was maintained by adding dilute $H_2SO_4$ before the addition of $ClO_2$.

## FTIR and XRD Analysis

Fourier Transformed Infrared Spectroscopy (FTIR) spectra of pulp samples were recorded with a resolution of 4 cm$^{-1}$ over the waves ranging from 4,000 to 400 cm$^{-1}$, using 32 scans per sample. FTIR spectrum was analyzed using the FTIR database.

Crystallinity of cellulose was determined for both untreated and enzymatically treated samples by X-ray diffraction (XRD). After freeze-drying, the samples were analyzed using X-ray diffractometer with an X-ray generator (from 2 to 50 of $2\theta$ scattering angle). The crystalline index of cellulose ($Xc$) was calculated from the X-ray diffraction patterns using the following equation:

$$Xc = \frac{(I_{002} - I_{am})}{I_{002}} \times 100$$

Where $I_{002}$ is the peak intensity from the (0 0 2) lattice plane ($2\theta$=22.3°) and $I_{am}$ is the peak intensity of amorphous phase ($2\theta$=15.7°).

## Analytical Studies

Compositional analysis of the agro-residual pulp after pretreatment with different enzyme samples was carried out according to the standard National Renewable Energy Laboratory (NREL) assay procedures [14]. The permanganate number and viscosity were determined according to Technical Association of Pulp and Paper (TAPPI) standards [15]. The brightness, whiteness, optical and color dimensions of hand sheets were also measured (L & W-070; Sweden). The strength of hand sheets was tested according to TAPPI test methods.

Mixed effluents from the different biobleaching strategies have been characterized by analyzing BOD and color following the American Society for Testing and Materials (ASTM) standards, ASTM D-1252-00, and ASTM-1209-00, respectively.

## Toxicity Analysis of Mixed Effluents

Mixed effluents ($C_D$, $E_{OP}$, $D_1$, and $D_2$ in the ratio of 3.25:1:1:1) from all the strategies were assessed using the 81.9% toxicity test protocol with a Microtox Model 500 Analyzer (Modern Water, USA). Different dilutions of each sample were prepared, and the osmolarity of each sample was adjusted with the osmolarity adjusting solution provided by the manufacturer. The endpoint measured by the Microtox assay detects the decrease in the intensity of light emitted by the luminescent marine bacterium *Vibrio fischeri* after 5 and 15 min of exposure. Hence, the $EC_{50\text{-}5min}$ and $EC_{50\text{-}15min}$ average values denote the effective concentration (EC) of effluent that caused a 50% reduction in the luminescence of the bacteria [16], [17]. These EC values were calculated using the standard 81.9% toxicity test for each collected effluent sub-sample to determine the acute toxicity of each collected sub-sample. Test results were entered into an Excel worksheet and corrected to normalize the color differences in different treated samples. The unit of acute toxicity (TOU) after 15-min incubation was also calculated for each sample to determine the EC of effluent required to kill 50% of the model organism. The sensitivity of each freeze-dried bacterial sample was periodically checked in our laboratory using zinc sulfate ($ZnSO_4 \cdot 7H_2O$) as a reference substance.

# RESULTS

## Optimization of Enzyme Production

All the nutritional and physical parameters affecting the xylanase production were screened through Plackett-Burman (PB) designing. Based on the regression analysis data of PB design, four significant factors [peptone ($X_1$), KNO3 ($X_2$), temperature ($X_3$) and incubation time ($X_4$)] were selected and studied at five different levels to analyze their individual and cumulative effect on xylanase production through CCD-RSM. An enhanced (4,220 U/g-dry cell weight) xylanase yield with a specific activity indicative of 597 U/mg-protein after 4 days incubation at 37°C under SSF conditions was achieved for *B. stearothermophilus* SDX by using two-step statistical modeling. The crude extract was free from cellulase and laccase activities and showed very low (12 U/mg-protein) amylase activity.

During laccase production under SmF by *C. subvermispora*, all the parameters were optimized using the "one variable at a time" approach. The highest specific activity of laccase in the crude extract was indicative of 10 U/mg-protein after 7 days incubation at 35°C and pH 5.0 under shaking conditions (50±5 rotations per minute) using 2.5% (w/v) wheat bran (Fig. 1). Crude extract of *C. subvermispora* was free from any cellulase and amylase activity, and contained very less (1.8 U/mg-protein) xylanase activity (Fig. 1). The commercial samples of xylanase and laccase showed specific activities indicative of 498 and 7 U/mg-protein, respectively.

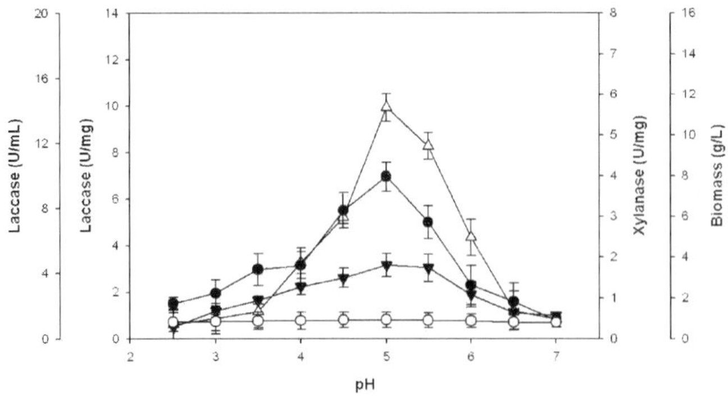


**Figure 1:** Effect of pH on laccase and xylanase production and growth of fungal mycelium of *C. subvermispora.*

*Experiments were carried out at 35°C for 7 days using 2.5% (w/v) of wheat bran in a 250-mL Erlenmeyer flask containing 50 mL of production media. Laccase activity (•); specific activity (~); xylanase activity (▼); fungal biomass (○).

https://doi.org/10.1371/journal.pone.0102581.g001

## Selection of Natural Mediator

*C. subvermispora* and commercial laccase were analyzed with Azure-B dye to verify the mediator capability of all the phenolic compounds. Decolorization of Azure-B was used to verify the potential of natural phenols to behave like mediators in lignin degradation [18]. Among all the tested natural mediators (AS, CA, PCA, DMP, DP, HBA, 4-HA, and SA), only SA showed a significant potential relative to other mediators. In comparison with HBT, AS showed a weak potential, which was reflected by only 5% decolorization.

## Parametric Optimization of Enzymatic Pretreatment

The operational parameters for xylanase-aided pretreatment were optimized through CCD-RSM. A second-order polynomial equation was used to analyze the individual and cumulative effects of four independent variables on the pulp properties (Table 1). Five different levels of each independent variable were used to analyze the regression of dependent variables in order to obtain the corresponding responses $Y_1$, $Y_2$, and $Y_3$ respectively and facilitate their statistical analysis. On the basis of the coefficient of determination ($R^2$=0.9029) and an adequate precision (6.446) for the model, 7.5 U of *B. stearothermophilus* xylanase at 60°C and pH 9.0 for 150 min of retention time (i.e., when all the independent variables were at their central point "0") was found to be the most effective for the pretreatment (Table 1). At optimal conditions, the viscosity of the xylanase produced in the lab and commercial xylanase pretreated pulps were 4.46±0.98 and 4.24±0.84 poise, respectively.

A similar strategy was applied to study the individual and cumulative effects of the laccase dose ($ED_L$), natural mediator concentration ($MC_L$), and retention time ($RT_L$) in the response of the dependent variables of the agro-residual pulp pretreatment (Table 2). Responses of the dependent variables were studied using the second-order polynomial equations. The value of $R^2$ indicates that a variation of only 5.41% is possible in the values of the dependent variables. From the statistical modeling it was observed that 20 U (coded value "0") of *C. subvermispora* laccase with 1.5% of SA concentration (coded value "0") was the most effective for an incubation period of 225 min (coded value "+α", Table 2). Other physical parameters (viz. temperature and pH) were optimized by using the "one variable at a time" approach and it was observed that the most suitable conditions for the LMS corresponded to 50°C and pH 5.0. After sequential pretreatment of the pulp, the viscosity of the enzyme produced in the lab and commercial enzyme pretreated pulps were 4.17±0.76 and 3.89±0.45 poise, respectively. Enzyme supplementation significantly reduced the viscosity of the pretreated pulps compared to the untreated pulp (5.18±0.96 poise).

The optimized physical parameters at which *B. stearothermophilus* xylanase and *C. subvermispora* laccase are the most effective were used for strategies III and IV, involving commercial enzymes. Pretreated agro-residual pulp was further characterized for compositional analysis. Under optimized conditions, the combination of *B. stearothermophilus* xylanase and *C. subvermispora* laccase along with SA was found to be the most effective among all combinations for removal of interfering residual materials.

## Residual Laccase Activity

Both enzyme samples were highly unstable in the presence of SA (natural mediator) as 94.5% and 82.5% of their activity was lost for *C.*

*subvermispora* and commercial laccase, respectively (Fig. 2). However, the presence of pulp stabilized the enzyme samples leading to only 34.8% and 16.3% loss of activity for *C. subvermispora* and commercial laccase, respectively. The pulp in the reaction mixture efficiently prevented denaturation of the enzymes even after the addition of SA to the reaction mixture, resulting in 42.8% and 28.8% loss of activity for *C. subvermispora*and commercial laccase, respectively. A surfactant (Tween-20, Tween-80 and Triton X-100) and a preservative (glycine) were also used to improve the residual activity of *C. subvermispora*laccase. Addition of Tween-20 (1.5%; v/v) provided extra stability to the *C. subvermispora*laccase leading to only 34.4% loss of activity.

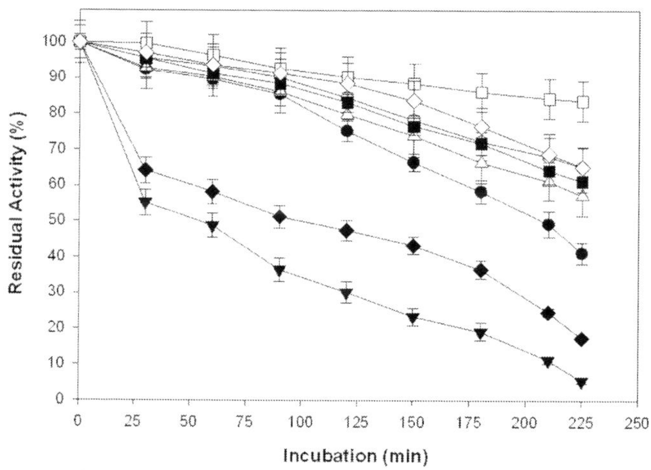


**Figure 2:** Effect of natural mediator and pulp on the residual activity of laccase during delignification.

(•) *C. subvermispora* only; (○) *C. subvermispora* with pulp only; (▾) *C. subvermispora*with SA only; (▵) *C. subvermispora* with pulp & SA; (▪) Commercial laccase only; (□) Commercial laccase with pulp only; (□) Commercial laccase with SA only; (◇) Commercial laccase with pulp & SA.

https://doi.org/10.1371/journal.pone.0102581.g002

## Development of the Enzymatic Process

With an optimized $ED_X$ of 7.5 U per gram of OD pulp at pH 9.0, strategy I, which involved the use of *B. stearothermophilus* xylanase, resulted in a 20.6% reduction in the consumption of $Cl_2$to give brightness identical to that observed for the conventional bleaching process. Strategy II ($ED_L$=20 U at pH 5.0 and 50°C), which involves the sequential use of *C. subvermispora*laccase resulted in a 30.2% reduction in the consumption of $Cl_2$ (Table 3). In order to compare the biobleaching efficiency of *B. stearothermophilus* SDX xylanase (strategy I), commercial Novozyme xylanase (Pulpzyme VLBL) was also used in strategy III, and it resulted in only 17.8% reduction in consumption of $Cl_2$. On the other hand, when commercial xylanase and laccase were used sequentially in strategy IV, a reduction of 25.3% in $Cl_2$ consumption was observed (Table 3).

**Table 3:** Different schemes for biobleaching agro-residual pulp with a reduced dose of chlorine*

https://doi.org/10.1371/journal.pone.0102581.t003

| Parameters | Control | Strategy - I | Strategy - II | Strategy - III | Strategy - IV |
|---|---|---|---|---|---|
| | 100%[a] | 79.4%[b] | 69.8%[c] | 82.2%[d] | 74.7%[e] |
| **After enzymatic pretreatment stage (X/XL)** | | | | | |
| Brightness (%ISO) | 43.6 | 45.8 | 46.9 | 45.2 | 46.4 |
| Yellowness | 15.9 | 13.9 | 10.9 | 14.4 | 11.7 |
| Whiteness | 23.5 | 24.1 | 25.4 | 23.8 | 24.6 |
| Permanganate no | 7.1 | 6.5 | 5.3 | 6.9 | 5.6 |
| Kappa no | 11.5 | 10.8 | 9.9 | 11.2 | 10.4 |
| **Chlorine, chlorine dioxide treatment stage ($C_D$)** | | | | | |
| Total $Cl_2$ added (%) | 2.50 | 1.99 | 1.75 | 2.06 | 1.87 |
| $Cl_2$:$ClO_2$ | 90:10 | 90:10 | 90:10 | 90:10 | 90:10 |
| $Cl_2$ Consumed (%) | 2.48 | 1.62 | 1.52 | 1.88 | 1.61 |
| pH | 2.5 | 2.5 | 2.5 | 2.5 | 2.5 |
| Time (min) | 45 | 45 | 45 | 45 | 45 |
| Brightness (%ISO) | 53.6 | 53.6 | 57.2 | 52.7 | 55.9 |
| Yellowness | 11.6 | 10.6 | 9.95 | 11.2 | 10.4 |
| Whiteness | 8.06 | 8.41 | 8.82 | 8.19 | 8.27 |
| **Alkali stage ($E_{OP}$)** | | | | | |
| Alkali added (%) | 1.8 | 1.8 | 1.8 | 1.8 | 1.8 |
| Time (min) | 120 | 120 | 120 | 120 | 120 |
| Temp (°C) | 70 | 70 | 70 | 70 | 70 |
| pH | 12 | 12 | 12 | 12 | 12 |
| Brightness (%ISO) | 58.4 | 58.4 | 59.6 | 58.2 | 59.7 |
| Yellowness | 9.15 | 8.85 | 6.95 | 9.07 | 7.92 |
| Whiteness | 17.0 | 18.6 | 19.5 | 17.8 | 18.8 |
| **$D_1$ stage** | | | | | |
| $ClO_2$ added (%) | 0.9 | 0.678 | 0.628 | 0.739 | 0.672 |
| $ClO_2$ consumed (%) | 0.806 | 0.552 | 0.504 | 0.627 | 0.517 |
| pH | 3.41 | 3.25 | 3.18 | 3.42 | 3.35 |
| Time (min) | 180 | 180 | 180 | 180 | 180 |
| Temp (°C) | 70 | 70 | 70 | 70 | 70 |
| Brightness (%ISO) | 84.5 | 81.6 | 88.8 | 80.2 | 86.1 |
| Yellowness | 2.95 | 2.66 | 2.05 | 2.77 | 2.12 |
| Whiteness | 71.5 | 75.9 | 73.9 | 71.6 | 70.9 |
| **$D_2$ stage** | | | | | |
| $ClO_2$ added (%) | 0.4 | 0.301 | 0.279 | 0.328 | 0.298 |
| $ClO_2$ consumed (%) | 0.317 | 0.263 | 0.191 | 0.313 | 0.211 |
| pH | 3.85 | 3.56 | 3.37 | 3.54 | 3.65 |
| Time (min) | 180 | 180 | 180 | 180 | 180 |
| Temp (°C) | 70 | 70 | 70 | 70 | 70 |
| Brightness (%ISO) | 88.1 | 88.8 | 89.6 | 88.6 | 88.4 |
| Yellowness | 1.42 | 1.27 | 1.01 | 1.40 | 1.11 |
| Whiteness | 81.6 | 82.5 | 83.2 | 82.1 | 82.8 |
| **$SO_2$ stage** | | | | | |
| Brightness (%ISO) | 88.9 | 89.4 | 89.9 | 89.1 | 89.2 |

*Volume of elemental chlorine ($Cl_2$) used in control process was considered as 100%.
a: Pulp without pretreatment with 100% $Cl_2$; b: Pretreated (with *B. stearothermophilus* xylanase) pulp having 75.4% of $Cl_2$; c: Pretreated (with *B. stearothermophilus* xylanase and *C. subvermispora* laccase) pulp having 69.8% of $Cl_2$; d: Pretreated (with commercial xylanase) pulp having 82.2% of $Cl_2$; e: Pretreated (with commercial xylanase and laccase) pulp having 74.7% of $Cl_2$.
doi:10.1371/journal.pone.0102581.t003

After the chlorination stage, an alkali has to be added to the reaction mixture before the addition of $ClO_2$. The main purpose of alkali addition is to eliminate the loosened and degraded lignin [19]. The addition of $ClO_2$ (0.9%) was performed in two steps. An excessive dose of $ClO_2$ or the addition of the entire volume of $ClO_2$ in a single step may reduce the strength of the pulp fiber [20]. Finally, sulfur dioxide ($SO_2$) was added to neutralize the remaining $ClO_2$ in the reaction mixture since residual $ClO_2$ might otherwise interfere with the final brightness [21]. By following the $XC_DE_{OP}D_1D_2$ bleaching sequence, an increase of 39.8% and 37.8% in the brightness was achieved after the $D_1$ stage for strategies I and III, respectively. On the other hand, the $XLC_DE_{OP}D_1D_2$ bleaching sequence resulted in a 48.9% and 44.2% increase in brightness after the same $D_1$ stage for strategy II and IV, respectively.

For the analysis of the strength of biobleached pulp, hand sheets (70±1 $gm^{-2}$) were prepared and acclimatized overnight under standard temperature and humidity conditions [22]. Among all the processes, strategy II was the most efficient in improving the strength, while strategy IV was the next most efficient (Table 4). Increases of 13.6% and 12.8% were observed in the tear index (mN $m^2/g$) in strategies II and IV, respectively. The improvements in other physical properties such as breaking length and porosity were also observed.

**Table 4:** Effect of enzymatic treatment on different physical properties of the agro-based pulp

https://doi.org/10.1371/journal.pone.0102581.t004

| Properties | Control | Strategy-I | Strategy-II | Strategy-III | Strategy-IV |
|---|---|---|---|---|---|
| Burst factor | 38.1±3.9 | 40.2±4.4 | 41.4±4.6 | 39.8±4.1 | 41.2±4.3 |
| Burst index (kN/g) | 3.72±0.42 | 3.94±0.41 | 4.06±0.42 | 3.90±0.41 | 4.04±0.47 |
| Tear factor | 47.1±4.3 | 51.8±5.3 | 52.6±5.5 | 51.3±4.9 | 52.2±4.7 |
| Tear index (mNm²/g) | 4.62±0.61 | 5.07±0.62 | 5.15±0.72 | 5.03±0.32 | 5.11±0.72 |
| Double fold (no.) | 65.0±8.1 | 74.0±10.1 | 76.1±11.2 | 72.2±8.1 | 76.0±9.1 |
| Tensile strength (Nm/g) | 55.3±5.8 | 57.2±6.1 | 58.1±5.35 | 57.1±5.8 | 57.9±6.2 |
| Breaking length (m) | 5640±230 | 5830±380 | 5920±420 | 5820±380 | 5910±440 |
| Porosity (sec/100 ml) | 86.4±8.3 | 88.3±7.8 | 89.1±7.9 | 87.9±8.4 | 88.9±9.1 |

doi:10.1371/journal.pone.0102581.t004

## X-ray Diffraction (XRD) Analysis

The X-ray diffraction analysis of control and enzymatically pretreated wheat straw pulp is shown. In all the control and enzymatically pretreated samples, the peak intensity from (002) lattice plane (2θ=22.3°)

represented the crystalline cellulose, while peak intensity of amorphous phase occurred at 2θ=15.7° [23]. The $I_{002}$ intensities (in cm) of samples pretreated with xylanase produced in the lab and commercial xylanase at 2θ=22.3° were 5.83 and 5.80, respectively. For sequentially pretreated substrate at 2θ=22.3°, the intensities were 9.90 and 9.16 for lab enzymes and commercial enzymes. Similarly $I_{am}$ intensities (in cm) at 2θ=15.7° are given for individual enzyme pretreated and sequentially pretreated biomasses. XRD analysis revealed that all of the strategies, i.e. individual enzyme pretreatment or sequential enzymatic pretreatment increased the crystallinity index of the pulp; however, the highest improvement in crystallinity was observed in the pretreatment strategy involving xylanase and laccase produced in the lab up to 68.6%.

## FTIR Spectral Analysis

FTIR spectra for single enzyme-pretreated and sequentially pretreated pulp samples showed several characteristic and prominent changes. The sequentially pretreated pulp showed absorption structures similar to single enzyme pretreated strategy but with different intensities. The main bands around the region 3,420–3,460 $cm^{-1}$depict –OH stretching of hydrogen bonding. The increase in their relative intensity after enzymatic treatments is attributed to the increase in cellulosic content of the pulp [24]. The variation in the relative intensity of bands at 2,921–2,917 $cm^{-1}$ and the decrease in the intensity of bands at 607–617 $cm^{-1}$, assigned to CH asymmetrical stretching vibration in $CH_3$, $CH_2$, and CH, in sequentially pretreated pulp indicated the degradation of aliphatic side chains. The relative intensity around 1,617–1,652 $cm^{-1}$ increased in sequentially pretreated pulp, which is attributed to the release of free carbonyl groups (C=O) due to the action of the enzyme on lignin's aromatic ring. In single enzyme pretreated pulp, these carbonyl groups remain associated with aromatic rings resulting in lesser absorption. The relative intensity of sequentially pretreated pulp decreased at 1,436–1,416 $cm^{-1}$, a band assigned to aromatic skeletal vibrations combined with $-CH_3$ in-plane deformations, showing that some methoxy groups were removed during the enzymatic treatment. The band at 1,320–1,382 $cm^{-1}$ was assigned to aliphatic C–H stretching in $CH_3$ (not in $-OCH_3$) and phenol-OH group. The decrease in its relative intensity illustrated that either the side chains or phen–OH of lignin decreased after the enzymatic treatment. The decrease in relative intensity at 1,266–1,252 $cm^{-1}$ indicates the degradation of guaiacyl groups. A new band that appeared at 1,737 $cm^{-1}$ in pulp pretreated with lab produced xylanase was assigned to C=O stretching vibration in β-C=O, COOH, and ester, indicating that residual lignin after an enzymatic pretreatment was enriched in these types of functional groups because of the removal of xylan. The changes in pulp crystallinity were derived from A1430/A897 ratio in accordance with the X-ray diffraction studies [24].

## Characterization of Effluents

Strategies II and IV resulted in 60.1% and 47.9% reduction in BOD, respectively, compared to the control process. These results were further supported by the color reduction values. The highest (25.8%) and lowest (12.1%) color reductions were observed with strategies II and III, respectively. The observed reduction in the color of the effluent was higher than that reported previously, where 11% reduction was reported with Pulpzyme HC [25]. Similarly, 60.1% reduction in BOD was observed with strategy II compared to 23.0% reduction with Kraft pulp in the previous report [26].

## Acute Toxicity Assay

The chemical intensive control process significantly inhibited the bioluminescence of *Vibrio fischeri* (Fig. 3a, b), as shown by the lower $EC_{50-5min}$ and $EC_{50-15min}$ values (Table 5). However, strategy I using *B. stearothermophilus* SDX xylanase more significantly reduced the acute toxicity of the mixed effluent (Fig. 3c, d), as shown by $EC_{50}$ values that were more than double the $EC_{50}$ of the control process (Table 5). Biological processes using Pulpzyme VLBL also reduced the acute toxicity (Fig. 3e, f); however, it was less efficient than the enzyme system produced in the lab (Table 5). Similar TOU (%) results were observed with these enzymes. A 2-fold improvement in TOU (%) was observed for strategy I compared to the control process, whereas the commercial xylanase resulted in 53.6% improvement in TOU (%). A slight increase in acute toxicity was observed with strategies involving the sequential application of xylanase and laccase compared to the single enzyme systems. In fact, each of these strategies alone, i.e., strategy II and IV, was significantly better than the control process in terms of the

$EC_{50}$ and TOU (%) values (Table 5). Sequential application of xylanase and laccase produced in lab, reduced the acute toxicity by 43.4–50.0% (Fig. 3 g, h) compared to the control process at different sampling times. In contrast, strategies involving commercial enzymes reduced the acute toxicity by 32.7–38.4% (Fig. 3i, j) compared to the control process at different sampling times (Table 5).

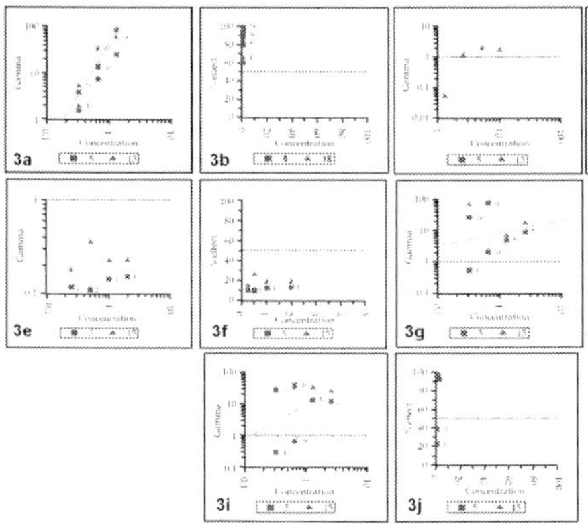


**Figure 3:** Gamma effect and $EC_{50}$ **values for (a, b) control; (c, d) strategy I; (e, f) strategy II; (g, h) strategy III;** (i, j) strategy IV.

The EC calculated by log linear plotting of Concentration (C) vs. % Light Decrease (%Δ), or more precisely by plotting Gamma (Γ) (which is the corrected ratio of the amount of light lost to the amount of light remaining) versus Concentration.

https://doi.org/10.1371/journal.pone.0102581.g003

**Table 5:** Reduction in acute toxicity of effluent samples determined through Microtox analyzer and represented by the average values of $EC_{50}$ **and unit of toxicity (TOU)***

https://doi.org/10.1371/journal.pone.0102581.t005

| Sample | Parameter | $EC_{50}$ (%) | TOU (%) |
|---|---|---|---|
| Control | $EC_{50\text{-}5min}$ | 39±4 | 221±26 |
| | $EC_{50\text{-}15min}$ | 32±3 | |
| Strategy I | $EC_{50\text{-}5min}$ | 84±9 | 119±14 |
| | $EC_{50\text{-}15min}$ | 82±9 | |
| Strategy II | $EC_{50\text{-}5min}$ | 69±7 | 152±19 |
| | $EC_{50\text{-}15min}$ | 64±7 | |
| Strategy III | $EC_{50\text{-}5min}$ | 72±7 | 134±16 |
| | $EC_{50\text{-}15min}$ | 69±8 | |
| Strategy IV | $EC_{50\text{-}5min}$ | 58±6 | 166±21 |
| | $EC_{50\text{-}15min}$ | 52±6 | |

*$EC_{50\text{-}5min}$: determination of inhibition in bioluminescence after 5 min of incubation; $EC_{50\text{-}15min}$: determination of inhibition in bioluminescence after 15 min of incubation; TOU (%) determination of toxicity unit after 15 min of incubation.
doi:10.1371/journal.pone.0102581.t005

# DISCUSSION

## Model for Pulp Pretreatment

Owing to the differences in the optimal pH values for the two enzymes and the inhibition of the activity of the mediator system under alkaline conditions, the pulp was subjected to sequential application of xylanase and laccase enzymes. Pretreatment was carried out with xylanase first because the mediators used during laccase pretreatment might generate free radicals capable of hindering the hydrolytic activity of xylanase [18]. Three-dimensional contour plots were also drawn, and it was observed that in xylanase-aided pretreatment, significant decrease in permanganate number was observed. On the other hand, a significant increase in the brightness was observed after the sequential application of enzymes. This might be because all the inhibitory components, which were generated after the use of individual enzymatic pretreatment, were neutralized in the sequential treatment process. Although numerous reports are available on the sequential use of xylanase and laccase for pulp pretreatment [3], [18], this is the first report on the sequential application of xylanase and laccase for biobleaching of the agro-residual pulp.

## Natural Mediator and Laccase Pretreatment

SA was selected as the mediator because of its plant origin and its higher efficiency to decolorize the Azure-B. Further, the improvement observed in the residual activity of the enzyme in the presence of pulp could be due to the cellulosic content of the pulp. Pulp fibers might provide an additional substrate for free radicals generated by the action of the natural mediator. The rate at which these free radicals target the enzyme is decreased owing to the presence of the pulp fiber; therefore, the residual enzyme activity increased. A similar observation was also reported by Fillat et al. [27]. Use of natural mediators will be beneficial in pollution reduction because the higher concentration of free radicals generated due to the action of synthetic mediators that make the whole bleaching process highly toxic and raise environmental concerns [19], [28]. Since commercial laccase has lower specific activity, it was less effective for delignification compared to *C. subvermispora* laccase, which has higher specific activity.

## Properties of Pulp obtained from Enzymatic Pretreatment Strategies

Improvement in the strength after enzymatic pretreatment is directly related to the hydrolytic action of xylanase. Hydrolytic enzymes disrupt the surface of cellulosic fiber and generate microfibrils, which in turn lead to a cross-networked, condensed packing of pulp fibers and hence they give extra strength [22]. Since only xylanase has hydrolytic properties, even sequential application of oxidizing enzymes (strategies II and IV) did not improve the strength significantly. Although commercial xylanase alone (Ecopulp, Cartazyme NS-10, and Pulpzyme HC) has been used previously [20], no report is available on the sequential application of xylanase and laccase for processing wheat straw pulp.

Improvement in crystallinity means a decrease in amorphous cellulose and increase in crystalline cellulose of the pulp. The increase in crystallinity might be due to the removal of hemicelluloses and lignin and components adhered to lignin as a result of sequential pretreatment, thereby increasing the cellulose content of the pulp [24]. Sequential pretreatment strategies involve separate laccase supplementation, and this is probably the reason for higher increase in crystallinity in sequential approach than in single enzyme strategy.

## Reduction in Pollution Load

Achieving less or non-toxic discharge from pulp and paper industry is the biggest challenge in the current scenario. Specific characteristics of the effluents such as BOD and color were determined, and strategy II was found to be the most effective in reducing the pollution load (color), resulting in 25.8% reduction. This might be because of the sequential use of *B. stearothermophilus* SDX xylanase and *C. subvermispora* laccase that removes a significant amount of lignin during pretreatment, which is further removed in the subsequent washes. Washing away the oxidized ligneous material and other hemicellulosic waste reduces the consumption of oxidizing chemicals during bleaching, thereby resulting in effluents with less BOD and colored units. Lesser pollution reduction efficiency of strategy III and IV could be attributed to the presence of surfactants and preservatives in commercial enzyme samples [20]. Media components present in crude enzymes produced in the lab do not interfere with the catalytic properties of the enzymes and hence do not alter their pollution reduction efficiencies [29]. The use of natural mediators in sequential enzymatic approach resulted in better performance in terms of reduction in pollution load. Therefore, the costs associated with the use of natural mediators can be compensated.

## Reduction in Acute Toxicity

The main reason for the reduced acute toxicity observed with the biological processes was the reduced consumption of oxidizing materials during bleaching. However, a variation was observed in the efficiency of the enzymes produced in lab and commercial enzyme systems. This difference in the efficiency could be explained by the presence or absence of chelators and preservatives. Commercial enzymes contain unidentified preservatives and other compounds to preserve their activity for an extended period of time, and these compounds have been shown to be crucial during the determination of acute toxicity and units of toxicity (http://www.brenntagspecialties. com/en/downloads/Products/Food/Novozymes/Cellulast_1_5L.pdf). Application of commercial enzymes liberates these compounds in the effluent and subsequently increases the biological load and hence the acute toxicity (http://www.sfm.state.or.us/cr2k_subdb/MSDS/CELLUCLAST. PDF). This could explain the lower toxicity of effluents generated using enzymes produced in lab.

Although the sequential application of xylanase and laccase efficiently removes the ligneous material during pretreatment, augmentation of the reaction by mediators caused an increase in the biological load of the effluent generated from these biological processes. Although the mediator was from a natural source, its addition still caused the formation of free radical compounds [13]. Once they formed, they immediately interact with other organic compounds to form highly stable toxicants, which increased the acute toxicity. Because of their natural origin, the concentration of the generated free radical compounds was much less than the intermediates and toxicants generated by the application of synthetic organic mediators [30]. Therefore, application of natural mediator is preferred over synthetic mediator.

## CONCLUSIONS

In the current study, sequential biobleaching of agro residual pulp has been proved to be a beneficial strategy compared to the conventional bleaching sequence. Significant reduction in the release of toxic organic compounds was achieved by incorporating sequential enzymatic pretreatment strategy with the conventional bleaching ($C_DE_{OP}D_1D_2$) process. Sequential pretreatment strategy combined with natural mediators further improved the eco-friendly nature of the process. Significant reduction in ecotoxicity also suggests the potential of this approach in the development of green technology for pulp and paper industry.

## AUTHOR CONTRIBUTIONS

Conceived and designed the experiments: SSD JS JKL. Performed the experiments: SSD GG. Analyzed the data: SSD JKL. Contributed reagents/materials/analysis tools: YCK. Wrote the paper: SSD VCK YCK JKL.

## REFERENCES

1. Garg G, Dhiman SS, Mahajan R, Kaur A, Sharma J (2011) Bleach-boosting effect of crude xylanase from *Bacillus stearothermophilus* SDX on wheat straw pulp. New Biotechnol 28: 58–64.
2. Yadav R, Chaudhary S, Dhiman SS (2010) Biopulping and its potential to reduce effluent loads from bleaching of hardwood kraft pulp. BioRes 5: 159–171.
3. Valls C, Vidal T, Roncero MB (2010) Boosting the effect of a laccase mediator system by using a xylanase stage in pulp bleaching. J Hazard Mat 177: 586–592.
4. Fillat U, Roncero MB (2009) Biobleaching of high quality pulps with laccase mediator system: influence of treatment time and oxygen supply. Biochem Eng J 4: 193–198.
5. Johannes C, Majcherczyk A (2000) Natural mediators in the oxidation of polycyclic aromatic hydrocarbons by laccase mediator systems. Appl Environ Microbiol 66: 524–528.

6. Dhiman SS, Sharma J, Battan B (2008) Pretreatment processing of fabrics by alkalothermophilic xylanase from *Bacillus stearothermophilus* SDX. Enz Microb Technol 43: 262–269.
7. Plackett RL, Burman JP (1946) The design of optimum multifunctional experiments. Biometals 33: 305–32.
8. Horikoshi K (1979) Production and industrial application of beta-cyclodextrin. Process Biochem 14: 26.
9. Bailey MJ, Biely P, Poutanen K (1992) Interlaboratory testing of methods for assay of xylanase activity. J Biotechnol 23: 257–270.
10. Miller GL (1959) Use of Dinitrosalicylic Acid Reagent for Determination of Reducing Sugar. Anal Chem 31: 426–428.
11. Bourbonnais R, Paice MG (1990) Oxidation of non-phenolic substrates. An expanded role for laccase in lignin biodegradation. FEBS Lett 267: 99–102.
12. Bradford MM (1976) A rapid and sensitive method for the quantitation of microgram quantities of protein utilizing the principle of protein-dye binding. Anal Biochem 72: 248–254.
13. Camarero S, Ibarra D, Martinez DT, Romero J, Gutierrez A, et al. (2007) Paper pulp delignification using laccase and natural mediators. Enzyme Microb Technol 40: 1264–1271.
14. Sluiter A, Hames B, Hyman D, Payne C, Ruiz R, et al. . (2008) Determination of structural carbohydrates and lignin in biomass. National Renewable Energy Laboratory (NREL), Golden Company.
15. TAPPI Test Methods (1996) Technical association of the pulp and paper industry. Atlanta, GA, TAPPI Press.
16. An J, Jeong S, Moon HS, Jho EH, Nam K (2012) Prediction of Cd and Pb toxicity to *Vibrio fischeri* using biotic ligand-based models in soil. J Hazard Mater 203–204: 69–76.
17. Valls C, Vidal T, Roncero MB (2010) Boosting the effect of a laccase-mediator system by using a xylanase stage in pulp bleaching. J Hazard Mater 177: 586–592.
18. Aracri E, Vidal T (2011) Xylanase- and laccase-aided hexenuronic acids and lignin removal from specialty sisal fibres. Carbo Poly 83: 1355–1362.

19. Ibarra D, Romero J, Martínez M, Martínez A, Camarero S (2006) Exploring the enzymatic parameters for optimal delignification of eucalypt pulp by laccase mediator. Enzyme Microb Technol 39: 319–1327.
20. Vicuna R, Escobar F, Osses M, Jara A (1997) Bleaching of eucalyptus Kraft pulp with commercial xylanases. Biotechnol Lett 19: 575–578.
21. Dhiman SS, Garg G, Mahajan R, Garg N, Sharma J (2009) 'Single lay out' and 'mixed lay out' enzymatic processes for bio-bleaching of kraft pulp. Bioresour Technol 100: 4736–4741.
22. Lundquist L, Willi F, Leterrier Y, Manson JAE (2004) Compression behavior of pulp fiber networks. Poly Eng Sci 44: 45–55.
23. Kaushik A, Singh M (2011) Isolation and characterization of cellulose nanofibrils from wheat straw using steam explosion coupled with high shear homogenization. Carbo Res 346: 76–85.
24. Virk AP, Puri M, Gupta V, Capalash N, Sharma P (2013) Combined enzymatic and physical deinking methodology for efficient eco-friendly recycling of old newsprint. PLOS ONE 8: 10. 137/journal. pone. 0072346.
25. Sudha B, Veeramani H, Sumathi S (2003) Bleaching of bagasse pulp with enzyme pre-treatment. Water Sci Technol 147: 163–168.
26. Yadav RD, Chaudhry S, Gupta S (2012) Novel application of fungal *Phanerochaete* sp. and xylanase for reduction in pollution load of paper mill effluent. J Environ Biol 33: 223–226.
27. Fillat A, Colom JF, Vidal T (2010) A new approach to the biobleaching of flax pulp with laccase using natural mediators. Bioresour Technol 101: 4104–4110.
28. Amann M (1997) The lignozym process coming closer to the mill. In: Nineth ISWFPC Protocol. Volume F4: pp. 1–5.
29. Lefebvre V, Vaes G (1989) The enzymatic evaluation of procollagenase and collagenase inhibitors in crude biological media. Biochem Biophys Acta 992(3): 355–361.
30. Camarero S, Ibarra D, Martinez MJ, Martinez AT (2005) Lignin-Derived compounds as efficient laccase mediators for decolorization of different types of recalcitrant dyes. Appl Environ Microbiol 71: 1775–1784.

CHAPTER 12

# CLONING, EXPRESSION AND CHARACTERISTICS OF A NOVEL ALKALISTABLE AND THERMOSTABLE XYLANASE ENCODING GENE (MXYL) RETRIEVED FROM COMPOST-SOIL METAGENOME


**Digvijay Verma[1], Yutaka Kawarabayasi[2,3], Kentaro Miyazaki[4,5], Tulasi Satyanarayana[1]**

[1] Department of Microbiology, University of Delhi South Campus, New Delhi, India

[2] Laboratory for Functional Genomics of Extremophiles, Faculty of Agriculture, Kyushu University, Fukuoka, Japan

[3] Institute for Hearth Research, National Institute of Advanced Industrial Science and Technology (AIST), Hyogo, Japan

[4] Bioproduction Research Institute, National Institute of Advanced Industrial Science and Technology (AIST), Hokkaido, Japan

[5] Department of Medical Genome Sciences, Graduate School of Frontier Sciences, The University of Tokyo, Hokkaido, Japan

# ABSTRACT


## Background

The alkalistable and thermostable xylanases are in high demand for pulp bleaching in paper industry and generating xylooligosaccharides by hydrolyzing xylan component of agro-residues. The compost-soil samples, one of the hot environments, are expected to be a rich source of microbes with thermostable enzymes.

## Methodology/Principal Findings

Metagenomic DNA from hot environmental samples could be a rich source of novel biocatalysts. While screening metagenomic library constructed from DNA extracted from the compost-soil in the p18GFP vector, a clone (TSDV-MX1) was detected that exhibited clear zone of xylan hydrolysis on RBB xylan plate. The sequencing of 6.321 kb DNA insert and its BLAST analysis detected the presence of xylanase gene that comprised 1077 bp. The deduced protein sequence (358 amino acids) displayed homology with glycosyl hydrolase (GH) family 11 xylanases. The gene was subcloned into pET28a vector and expressed in *E. coli* BL21 (DE3). The recombinant xylanase (rMxyl) exhibited activity over a broad range of pH and temperature with optima at pH 9.0 and 80°C. The recombinant xylanase is highly thermostable having T1/2 of 2 h at 80°C and 15 min at 90°C.

## Conclusion/Significance

This is the first report on the retrieval of xylanase gene through metagenomic approach that encodes an enzyme with alkalistability and thermostability. The recombinant xylanase has a potential application in paper and pulp industry in pulp bleaching and generating xylooligosaccharides from the abundantly available agro-residues.


# INTRODUCTION

Hemicellulose represents the second most abundant renewable polymer of plant cell walls after cellulose, and xylan is the main constituent in lignocellulosic agro-residues. The backbone of xylan is mainly composed of β-1,4-linked xylosyl residues along with various groups (arabinosyl, acetyl and glucuronosyl) in their side chains. These heterogeneous polysaccharides play a critical role in maintaining the cell wall integrity by making covalent and non-covalent bonds with cellulosic fibres and lignins [1], [2]. The heterogeneity in xylans necessitates multiplicity in the xylanolytic enzyme system that comprises endoxylanase (1,4-β-D-xylan xylanohydrolase; EC 3.2.1.8), β xylosidase (1,4 β-D-xylan xylohydrolase; EC 3.2.1.37) α-glucuronidase, α-L-arabinofuranosidase, and acetyl xylan esterase. The CAZY database (http://www.cazy.org/fam/acc_GH.html) categorized xylanases into six glycosyl hydrolase families GH5, GH8, GH10, GH11, GH30 and GH43 [3]. Among these, family 10 and 11 xylanases are widely distributed where GH10 xylanases share versatile substrate specificity with higher molecular weight, while GH11 xylanases are more stringent in their substrate specificity with low molecular weight and are considered as true xylanases [4], [5]. Hydrolysis of xylans by xylanases is of interest to many industries like ramie fibre degumming, food processing, textile, biofuel, feed and paper/pulp industries [6], [7], [8]. The xylanases that withstand extreme conditions prevailing in the industrial processes are in high demand in paper pulp processing and feed industries.

Although several xylanases have been reported from a number of microorganisms, most of them do not have adequate thermostability and alkalistability for their application in paper and pulp industries. Majority of xylanases have been obtained from the culturable 0.1 to 1% of the total microbial diversity existing in natural environments. The culture-independent metagenomic approaches permit retrieval of genes encoding useful enzymes from environmental samples without involving laborious and elaborate methods for the cultivation of microbes. The immense demand for alkalistable and thermostable xylanases encouraged us to adapt this innovative strategy for retrieving genes that encode thermo-alkali-stable xylanases. In this investigation, a metagenomic library was constructed and screened for clones with xylanase activity. Xylanase encoding gene (*Mxyl*) was subcloned and expressed, and the recombinant xylanase was purified and characterized. To the best of our knowledge, this is the first report on retrieving thermo-alkali-stable xylanase by a metagenomic approach.

# RESULTS

## Construction of Metagenomic Library, DNA Sequencing and Bioinformatics Analysis

When 5.0 μg of 20–30 kb of high molecular weight metagenomic DNA was digested with Sau3AI and the fragments were ligated into p18GFP vector with an efficiency of $3.6\times10^4$ clones per μg of DNA in constructing the library, the insert sizes were in the range of 3.0–8.0 kb with an average size of 5.5 kb. On screening, a clone having xylanase gene was spotted on RBB xylan containing LB-amp plate. The full sequence of the insert showed the size of 6.231 kbp that revealed its prokaryotic origin on blast analysis. The complete insert contained nine transcriptional units with a complete ORF of 1077 bp long xylanase gene. The sequence showed putative sequences of -35 (CACGCCA), -10 (TAAAAA) and ribosomal binding sites (AGGGG) at the upstream of xylanase gene followed by complete ORF having ATG and TAA as start and stop codons, respectively. The xylanase displayed five conserved regions (I-V) of GH11 xylanase having two catalytically important residues ($Glu_{109}$ and $Glu_{217}$) present in signature sequence II and V. Amino acid homology showed maximum identity (79%) with the xylanase gene of an uncultured bacterium and *Actinoplanes* sp. SE50/110 followed by a metagenomic GH 11 xylanase (71%). It shared 63–75% homology with xylanases produced by various species of *Streptomyces*. The xylanase retrieved in this investigation exhibits 75, 67 and 64% similarity with the endo-1,4 β-xylanases of *Cellulomonas fimi*, *Micromonospora aurantiaca* 27029 and *Amycolatopsis mediterranei* U32, respectively. It, however, has lower homology with the xylanases of *Microbulbifer hydrolyticus* (63%), *Pseudomonas* sp. ND137 (62%), uncultured *Cellvibrio* sp. (58%), *Cellvibrio mixtus* (57%) and *Aspergillus fumigatus* AF293 (52%) (Figure 1).

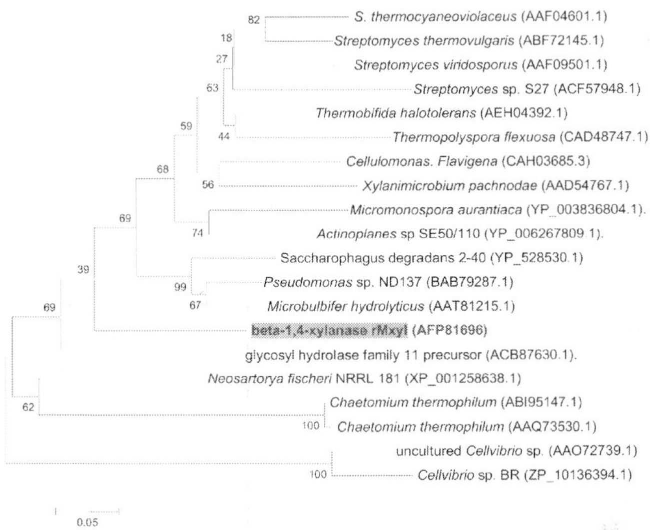


**Figure 1:** Phylogenetic tree of recombinant xylanase.

rMxyl showed highest homology with xylanase of *Cellulomonas fimi* ATCC 484 followed by uncultured microbial GH 11 xylanase. Neighbour Joining (NJ) Tree is constructed by using MEGA 4.0 software. Bootstrap values (n=1000 replicates) are represented as percentage. The scale bar depicts the allowed changes per amino acid position.

https://doi.org/10.1371/journal.pone.0052459.g001

## Construction and Expression of the Xylanase Gene in *E. coli* and Localization of the Encoding Recombinant Xylanase (rMxyl)

Xylanase gene was successfully cloned into pET28a(+) vector. It was confirmed by the release of insert from the vector on double digestion with NheI and XhoI. The sequencing results further confirmed the insert. The recombinant plasmid was expressed in *E. coli* BL21 (DE3) on induction with 1.0 mM IPTG at $A_{600}$ of 0.6–0.7 and 30°C. At higher level of expression, it led to the formation of inclusion bodies, which could be solubilized using 6.0 M urea. The highest titre of the recombinant enzyme was achieved in 4–6 h. The construct (*pET28a-Mxyl*) expressed a high proportion of xylanase in cytoplasmic fraction (83%), followed by periplasmic (9%) and extracellular (8%) fractions after 4–5 h of induction. When xylanase gene was cloned and expressed in pET22b(+) vector, a high proportion of intracellular enzyme (>60%) was produced in the initial 3 h of induction, and thereafter, it declined. The periplamic xylanase was optimum at 12 h, while the extracellular fraction gradually increased and it reached a peak (29%) in 24 h.

## Site Directed Mutagenesis

Muteins having $Glu_{117}$Asp and $Glu_{209}$Asp completely lost the activity. These two glutamates are highly conserved residues in the signature sequences LV**E**YYIVDN and MAT**E**GY, and these are responsible for catalytic activity of GH 11 xylanase.

## Purification, Biochemical Characterization and Zymogram Analysis of rMxyl

The recombinant xylanase was purified by $Ni^{2+}$-NTA resin affinity chromatography and the purified recombinant protein could be eluted using imidazole (100–400 mM). The protein appeared as a single band of 40 kDa against the protein marker on 15% SDS-PAGE, and the recombinant xylanase revealed as a clear band of xylan hydrolysis by zymogram analysis (Figure 2). The effects of various physical and chemical parameters on the recombinant xylanase have been assessed. The xylanase exhibited broad range of pH (6.0–12.0) with optimum at 9.0, and it retained ~55% residual activity at pH 10.0 (Figure 3A). The rMxyl is active in the temperature range between 40 and 100°C (Figure 3B) with optimum at 80°C, and retains more than 90–95% activity after exposure to 60 and 70°C for 3 h. The enzyme has a $T_{1/2}$ of 2.0 h at 80°C and 15 min at 90°C (Figure 3C). The recombinant enzyme did not lose activity after 3 h exposure to pH 8.0 and 9.0, and thereafter, it declined (50% residual activity after 4 h). Approximately 20–45% loss in activity was recorded on either side of the pH optimum after 1 h incubation (Figure 3D). $Mg^{2+}$, $Sn^{2+}$ and $Fe^{2+}$ stimulated rMxyl activity, while $Hg^{2+}$ and $Mn^{2+}$ significantly inhibited enzyme activity even at 1 mM. Other metal ions exerted inhibitory action on xylanase. More than 30% activity was lost in the presence of $Mn^{2+}$ (Table 1). NBS and PMSF inhibited the activity to a significant extent even at 1 mM concentration. β-ME and DTT strongly inhibited enzyme activity. A stimulatory effect EDTA was recorded on xylanase activity.

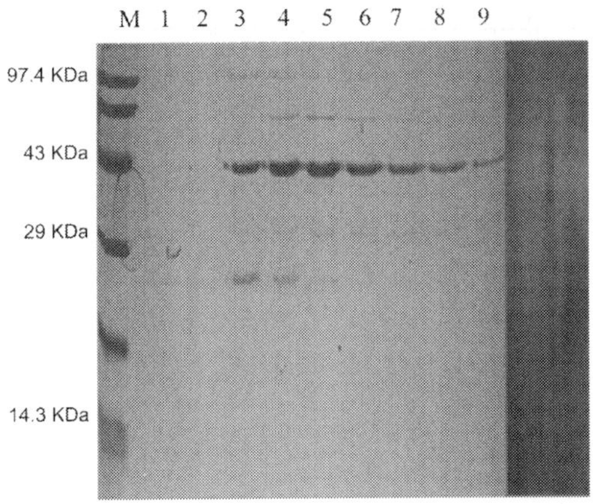


**Figure 2:** Analysis of rMxyl using SDS–PAGE (15% polyacrylamide gel).

A. Lane 1 protein marker, Lane 2 and 3 are washes with 20 and 30 mM mM imidazole. Recombinanat xylanase was eluted using different concentrations of imidazole (100, 200, 250, 300, 400, 450, 500 mM). Purified xylanase showed molecular mass of ~42 kDa on staining with Coomassie Brilliant Blue R-250. B. Zymogram analysis of purified xylanase using Congo red staining method.

https://doi.org/10.1371/journal.pone.0052459.g002

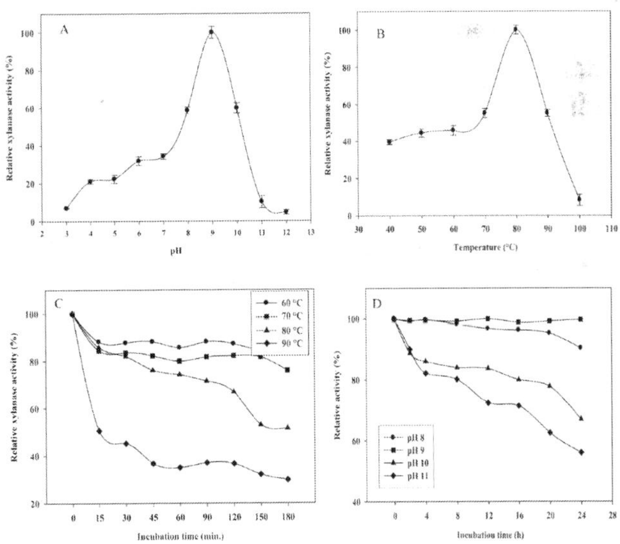


**Figure 3:** Effect of pH and temperature on the activity and stability of rMxyl.

A and B: The recombinant xylanase incubated in various buffers (pH 3–12) and temperatures (40–100°C) and assayed for xylanase activity C: Recombinant xylanase was incubated in glycine-NaOH buffer without substrate and kept at various temperatures. Aliquots were collected at various time interval and store at 0°C for calculating residual activity. D: Similarly enzyme was incubated in various buffers (pH 8–11) and aliquots of different time intervals were used xylanase assays. https://doi.org/10.1371/journal.pone.0052459.g003

**Table 1:** Effect of modulators on rMxyl activity

https://doi.org/10.1371/journal.pone.0052459.t001

| Metal ions | 1 mM | 5 mM | 10 mM |
|---|---|---|---|
| $Mg^{2+}$ | 106.45±1.05 | 99.65±0.98 | 87.38±0.45 |
| $Fe^{2+}$ | 108.65±0.75 | 116.01±0.27 | 93.67±1.32 |
| $Sn^{2+}$ | 110.43±0.67 | 76.12±0.44 | 45.17±0.63 |
| $Ni^{2+}$ | 91.21±0.22 | 79.01±1.34 | 32.84±0.43 |
| $Zn^{2+}$ | 91.67±0.32 | 76.64±0.78 | 32.89±0.89 |
| $Pb^{2+}$ | 81.33±067 | 20.78±0.32 | 09.65±0.67 |
| $K^{+}$ | 81.21±1.08 | 20.62±0.12 | 12.67 ±0.45 |
| $Ag^{2-}$ | 73.48±0.53 | 54.55±0.69 | 27.83±0.98 |
| $Ca^{2-}$ | 72.43±0.43 | 35.45±0.21 | 12.09±0.19 |
| $Mn^{2+}$ | 71.76±0.63 | 27.34±1.32 | 09.67±0.27 |
| $Ba^{2+}$ | 66.45±0.67 | 23.91±0.34 | 18.65±0.33 |
| $Cd^{2+}$ | 54.67±0.43 | 29.33±0.49 | 12.87±0.65 |
| $Co^{2+}$ | 59.15±1.23 | 29.63±0.65 | 12.54±1.12 |
| $Na^{+}$ | 61.43±0.78 | 39.75±1.06 | 27.35±0.78 |
| $Cu^{2+}$ | 29.12±0.18 | 15.76±0.76 | 10.09±0.87 |
| $Hg^{2+}$ | 0 | 0 | 0 |
| **Inhibitors** | **1 mM** | **5 mM** | **10 mM** |
| NBS | 46.66±0.12 | 35.67±0.09 | 20.12±0.11 |
| IAA | 103.45±0.54 | 89.75±0.32 | 69.85±1.56 |
| β-ME | 0 | 0 | 0 |
| DTT | 0 | 0 | 0 |
| EDTA | 105.65±1.23 | 107.19±1.01 | 89.98±0.56 |
| **Detergents** | **0.1% (v/v)** | **0.5% (v/v)** | |
| Tween 20 | 103.45±1.32 | 105.67±0.98 | |
| Triton X100 | 108.32±0.96 | 104.05±0.92 | |
| SDS | 97.34±1.32 | 65.89±0.19 | |
| Control | 100±0.12 | 100±0.23 | 100±0.67 |

Most of the metal ions were insignificant at 1 mM concentration; however xylanase activity was significantly inhibited at higher concentration by $Pb^{2+}$, $Ag^{2-}$, $Ca^{2+}$, $Mn^{2+}$, $Ba^{2+}$, $Cd^{2+}$ and $Co^{2-}$. In the presence of $Hg^{2+}$, enzyme lost completely its activity. Similarly, trace amounts of β-mercaptoethanol (β-ME) and dithiothreitol (DTT) completely inhibited the xylanase activity. Inhibition in the presence of N-bromosuccinimide (NBS) signifies the role of tryptophan in catalysis, while EDTA confirms it as a non-metallozyme.
doi:10.1371/journal.pone.0052459.t001

## Saccharification of Agroresidues/Hydrolysis of Xylan

The rMxyl hydrolyzed xylan from various sources. The enzyme activity was very high in birchwood xylan (relative activity 100%) in comparison with that on xylan from beech wood (97%) and arabinoxylan (80%). There was no activity on carboxymethylcellulose (CMC) and other non-xylan polysaccharides (starch, pullulan and chitin). The $K_m$ and $V_{max}$ values of the enzyme on birchwood xylan are 8.0±1.21 mg/ml and 300±09.12 μmol/min/mg, respectively. The saccharification of wheat bran was high (15.2%) as compared to that of corncobs (9.89%) and sugarcane bagasse (4.71%). Various xylooligosaccharides were detected in the hydrolysates (Figure 4).

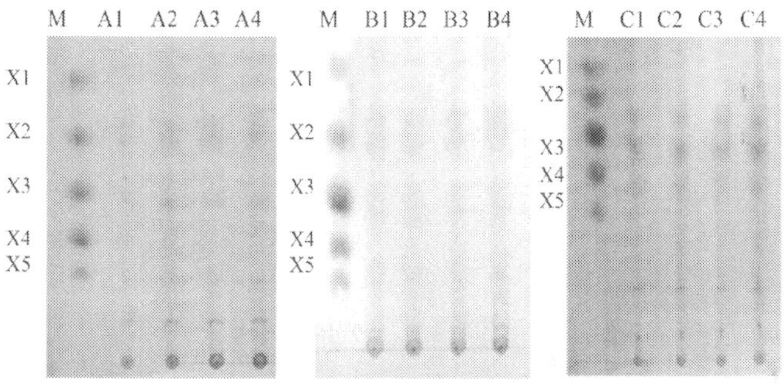


**Figure 4:** Profile of xylooligosaccharides liberated by the action of rMxyl.

Lane (A1–A4)*: spots of X1, X2 and X3 were detected from wheat bran. Lane (B1–B4)*: hydrolysis from corncob that showed prominence of X2 and X3. While, X3, X4 and X5 were detected from hydrolysate of sugarcane bagasse (C1–C4)*. Lane M: Standards of various XOs (Sigma, USA). X1: xylose, X2: xylobiose; X3: xylotriose; X4: xyloptetraose; X5: xylopentaose *: 1/2/3/4 time intervals of 5, 15 and 30 min and 1 h, respectively.

https://doi.org/10.1371/journal.pone.0052459.g004

# DISCUSSION

Although several xylanases have been reported from diverse microbiota using traditional culture dependent approaches [9], [10], [11], [12], majority of them do not endure the extreme temperature and alkaline conditions prevailing in industrial processes. An alternate strategy was, therefore, adapted to retrieve a thermo-alkali-stable xylanase gene (*Mxyl*) by culture-independent metagenomic approach. The metagenomic library constructed with DNA extracted from the compost-soil samples yielded a clone that produced xylanase. Although, the compost soils are in the acidic pH range, an alkalistable and thermostable endoglucanase had been reported from rice straw compost [13]. The culture independent approach has started yielding the useful biocatalysts from the hidden Pandora's Box [14], [15], [16]. A considerable success has also been achieved in obtaining xylanases with diverse attributes by using metagenomic approaches [17], [18], [19], [20], [21]. The protein encoded by xylanase gene comprises 358 amino acids, of which 16 are acidic and 21 basic.

The predicted molecular weight, pI and instability index of recombinant xylanase are ~40 kDa, 8.8 and 33.44 respectively. The xylanase contained a 43 amino acid long leader sequence at the N terminal region followed by a catalytic domain ($44^{th}$–$212^{th}$) of GH11 family interrupted by a short stretch of arginine and threonine rich non-catalytic region ($WSVRQ_2R_2TG_2TIT_2$). In addition, serine rich Q linker region ($S_2GS_2DITVG_2TS_2G_2TS_2G_2S_3G_2S_{10}G_4$) has also been detected from amino acid 213 to 248 just after catalytic domain. Such repeated amino acids make linker regions that usually discriminate catalytic domain from carbohydrate binding domain [22], [23], [24]. Moreover, linkers have also been reported as integral parts of various xylanases that connect thermo-stabilizing domains, surface layer homology domains and dockerin domains which play a role in stabilizing the protein [22], . Amino acid homology and hydrophobic cluster analysis categorized this high molecular weight xylanase into GH 11 family. This is not unusual because xylanases of *Clostridium stercorium* also had been classified as a GH11 xylanase despite its high molecular mass (56 kDa) [29]. Metagenomic origin, distinct characteristics, lower homology and higher molecular weight (>30 kDa) make this a novel xylanase. The recombinant xylanases from *Glacieola mesophila* KMM241 and *Geobacillus thermoleovorans* displayed a similar pattern of recombinant protein localization [30], [31]. The profile of the distribution of the recombinant xylanase is similar to other xylanases cloned in pET28a(+) and pET22b(+) vectors as in *G. thermoleovorans* and *B. halodurans* [30], [32]. The integrated N- terminal pelb signal sequence in pET22b(+)

directed the enzyme to periplasm that further led to secretion into the extracellular environment.

The site directed mutagenesis of two residues of glutamate to aspartate resulted in a complete loss of xylanase activity due to disruption in double displacement mechanism [33]. In order to take the advantage of thermostability of the recombinant xylanase, it was subjected to high temperature prior to purification by $Ni^{2+}$-NTA agarose resins. This step reduced the extra load of non-His tagged, less thermostable and contaminant host proteins [30], [32].

The rMxyl exhibits optimum activity at higher temperature and pH which is similar to xylanases produced by *Dictyoglomus thermolacticum*, *Thermotoga maritima*, *Bacillus acidicaldarius* and *Geobacillus thermoleovorans* having optimal activity at or above 80°C [30], [34], [35], [36]. The activity and stability of rMxyl at higher pH are the crucial properties of xylanases for their applicability in paper processing industry. Xylanases of *G. thermoleovorans*, *Bacillus firmus*, *B. stereothermophilus* T-6 exhibited optimum activity at pH at 8.0 or above [30], [37], [39]. The shelf life of rMxyl is more than 3 months at 4°C, which retains greater than 90% activity. The recombinant xylanase is optimally active at 80°C and pH 9.0 that distinguishes it from already reported xylanases. The xylanase of *Thermotoga maritima* has $T_{opt}$ of 90°C, but it gets inactivated fast at pH 6.0 [35]. The highly thermostable xylanase with optimum temperature in the range between 80 and 105°C are available, but these xylanases exhibit their maxima either at acidic or neutral pH [35,35]. Similarly the alkalistability at higher pH is reported in many xylanases but are active at lower temperatures [37], [38]. The recombinant xylanase of GH10 family from *Bacillus halodurans* showed both properties together having optima at 75°C and pH 9.0, but it losses 50% activity at 65°C after 4 h and gets inactivated very fast at 80°C [32]. The metagenomic xylanase, on the other hand, has good thermostability at higher temperatures (60, 70 and 80°C) with only 20–30% loss after 3 h exposure. The most significant aspect of this investigation is obtaining a highly alkalistable ($pH_{opt.}$ 9.0) and thermostable ($T_{opt.}$ 80°C) xylanase from environmental samples by a metagenomic approach.

Cations ($Mg^{2+}$, $Sn^{2+}$ and $Fe^{2+}$) stimulated the rMxyl activity like those of *B. subtilis* AMX-4 and an uncultured microbe [39], [40]. Even at 1 mM, $Hg^{2+}$ and $Mn^{2+}$ significantly inhibited the activity. The inhibition of xylanase by $Hg^{2+}$ suggests the presence of tryptophan residues that oxidize indole ring, thereby inhibiting the xylanase activity [39], [41], [42], [43]. Xylanases from *Streptomyces olivaceoviridis* A1, *Streptomyces* sp. S27 and *Bacillus subtilis* strain R5 had been reported to be stimulated by $Fe^{2+}$ and $Mg^{2+}$ [19], [44], [45], [46], [47], and total loss of the enzyme in presence of $Hg^{2+}$ and $Mn^{2+}$ was reported in the xylanases from an uncultured microbe [19], [46], *Penicillium* sp. [48] and *Plectosphaerella cucumerina* [41]. The inhibition of xylanase activity by $Cu^{2+}$ is similar to the majority of the xylanases [13], [39], [45], [49]. The stimulatory effect of EDTA could be due to a different catalytic mechanism as reported in xylanases of *Aspergillus niger* and *Fusarium proliferatum* [42], [49]. In *Glaciecola mesophila*KMM 241, EDTA caused ~25% enhancement in activity [31]. N-BS inhibition suggests the involvement of tryptophan in xylanase activity [50], [51], [52]. Total loss of xylanase activity by β-ME and DTT suggests the distortion of disulfide linkages present between cystein residues [41], [49]. Detergents exerted a slight stimulatory effect on the recombinant xylanase which is a common feature of the other xylanases. However, rMxyl was inhibited by SDS as in other xylanases [39], [46], [49].

The rMxyl hydrolyzed birchwood and beechwood xylan efficiently. The structural similarity of beechwood and birchwood xylans may be the reason for the high activity [31]. The enzyme exhibited almost similar activities on oat spelt and arabinoxylan. Oat spelt xylan is a type of arabinoxylan very rich in arabinose (xylose/arabinose =66:34) [53], [54]. Interestingly the rMxyl liberated xylooligosaccaharides from xylan in just 5 min and it was sustainable on prolonged incubation. Several xylanases have been reported from various microorganisms that liberate xylooligosaccharides following xylan hydrolysis. Alkaline xylanases show better action on agro-residues by lowering the steric hindrance caused by cellulose and enhancing the solubility of hemicellulosic materials [54]. Xylanases from *G. thermoleovorans* [30], *Bacillus halodurans* [55]*Thermomonospora fusca* [56] have been reported to generate a similar profile of XOs. The metagenomic xylanase finds application in food industry for the production of xylooligosaccharides as prebiotics [57].

# CONCLUSIONS

Most of the xylanases retrieved by culture-dependent and culture-independent approaches exhibit optimum activity in the pH and temperature range of 6.0–8.0 and 40–60°C, respectively. The xylanase obtained in this investigation through metagenomic approach (rMxyl) not only displays alkalistability and thermostability, but it also has a high thermostability. This is the first report of xylanase with twin stabilities obtained through culture-independent approach. A very low similarity in amino acid sequence of the enzyme with other known xylanases makes it a novel xylanase. The possibility of obtaining thermo-alkali-stable xylanase from composts may lead to an intense search for similar enzymes in this niche.

# MATERIALS AND METHODS

## Ethics Statement

No specific permits are required for collecting environmental samples used in this investigation. The samples did not have any animal or plant species because of the elevated temperature. The location is not privately owned or protected in any way and does not involve endangered and protected species.

## Sample Collection

The samples of compost-soil were collected in sterile polyethylene bags from the vicinity of a hot water spring near Fukuoka Japan and stored at 4°C. The pH of the samples is in the acidic range (3.0–4.5).

## Strains and Plasmids for DNA Manipulations

Cloning vector p18 GFP was the gift from Dr. Taku Uchiyama, Japan. The pGEM-T Easy vector (Promega, USA), pET28a(+) and pET22b(+) of Novagen (Germany) were used for sequencing and expression of the xylanase gene, respectively. *Escherichia coli* DH10B, *E. coli* XL1 blue and *E. coli* BL21 (DE3) were used for the propagation of the plasmid. Restriction enzymes and T4 DNA ligase were purchased from New England Biolabs (UK).

## Construction of Metagenomic Library

Soil DNA was extracted according to Verma and Satyanarayana (2011) [58]. Metagenomic DNA was processed for constructing the metagenomic library. Five µg of metagenomic DNA was partially digested with 0.5 U of restriction enzyme Sau3AI. The fragments of 3–12 Kb were eluted from agarose gel (1.2%, w/v) by gel extraction kit according to manufacturer›s protocol (Macherey-Nagel, Germany). Hundred nanogram (ng) of insert DNA and 300 ng of BamH I digested and dephosphorylated p18 GFP vector were ligated by using T4 DNA ligase overnight at 16°C. The ligation mixture was transformed into competent *E. coli* DH10B cells by heat shock method. The metagenomic library was spread and screened for xylanase activity on 0.3% (w/v) RBB-xylan (4-O-methyl-D-glucurono-D-xylan-remazol brilliant blue R) (Sigma, St. Louis, MO, USA) LB-ampicillin agar plates. The transformants were grown at 37°C overnight and observed for the zone of xylan hydrolysis.

## DNA Sequencing and Bioinformatics Analysis

The pure clone (TSDV-MX1) showing clear zone of xylanase hydrolysis was sequenced using M13 forward and reverse primers followed by different internal primers using Applied Biosystem 373 stretch automated sequencer (Applied Biosystems, Foster City, CA, USA) at Nucleic acid sequencing facility of the University of Delhi South Campus, New Delhi (India) for obtaining full sequence of the insert. The ORFs were identified by using the NCBI's open reading frame (ORF) Finder tool (http://www.ncbi.nlm.nih.gov/gorf/gorf.html). BLASTN and BLASTP of NCBI were used to align the nucleotide and amino acid sequences, respectively. Multiple alignments of the amino acids were carried out using the CLUSTALW programme (http://www.ebi.ac.uk/clustalW). The phylogenetic analysis was done using MEGA 2.1 with neighbour joining strategy.

## Construction of Expression Plasmids *pET28a-Mxyl* and *pET22b-Mxyl*

The full length xylanase gene was pulled out from the clone (TSDV-MX1) by PCR using specific primers MxylF1 (CCCGCTAGCATGACAGCGAGTTTGAGGAAGA) and

MxylR1 (CCCCTCGAGTTACGGCGTGTTTCCGTAGC) having the compatible NheI and XhoI restriction sites of pET28a(+) in the forward and reverse primers, respectively. Mxyl F2 (CCCG GATCCATGACAGCGAGTTTGAGGAAGA) and Mxyl R2 (CCCCTCGAGTTAC-GGCGTGTTTCCGTAGC) were used for cloning into pET22b(+) vector. The gene was amplified under the defined PCR conditions (initial denaturation 3 min at 95°C followed by 29 cycles of 10 sec at 98°C, 30 s at 59°C and 1 min at 68°C with a final extension step at 68 °C for 10 min). The PCR products pET28a(+) and pET22b(+) were digested with the restriction enzymes and ligated into the vectors followed by transformation into competent *E. coli* XL1 blue cells to obtain *pET28-Mxyl and pET22-Mxyl.*

## Expression of Xylanase (rMxyl) using *pET28a-Mxyl* and *pET 22b-Mxyl*

The recombinant constructs were confirmed by colony PCR followed by double digestion of the construct with restriction enzymes. These clones were processed for sequencing. The recombinant plasmid having the accurate sequence was then transformed into *E. coli* BL21 (DE3) competent cells for expression of the recombinant protein. The transformants were grown in kanamycin (50 µg/ml) containing LB medium at 37°C with 200 rpm in an incubator shaker to achieve an optical density of O.D. =0.5 to 0.7. Afterwards the expression was induced by adding isopropyl-β-D-1-thiogalactopyranoside (IPTG) to a final concentration of 1 mM and the culture was further cultivated at 30°C. The samples were collected at 1 h intervals for determining the enzyme titres.

## Site Directed Mutagenesis

Multiple sequence alignment of recombinant xylanase with those of the known xylanases revealed $Glu_{117}$ and $Glu_{209}$ as catalytically important residues. Experimentally it has been proved by site directed mutagenesis using Geneart site directed mutagenesis kit (Invitrogen, Carsband, USA). Two point mutations ($Glu_{117}$Asp and $Glu_{209}$Asp) were created in the native xylanase gene and expressed in *E. coli* BL21 (DE3). The induced mutations were confirmed by sequencing.

## Localization of rMxyl

Localization of the recombinant protein was determined by collecting the intracellular, extracellular and periplasmic fractions from the same culture. Extracellular, intracellular and periplasmic fractions were collected according to Verma and Satynarayana, 2012 [30].

## Xylanase Assay

Xylanase was assayed according to Archana and Satyanarayana (1997) [59]. The 0.5 ml of appropriately diluted enzyme was incubated at 80°C with 0.5 ml of 1% (w/v) birchwood xylan (Sigma, St. Louis, Mo.) in 100 mM glycine-NaOH buffer (pH 9.0) for 10 min. The reducing sugars were determined using 3, 5-dinitrosalicylic acid (DNSA) reagent [60]. One unit of xylanase is defined as the amount of enzyme required to liberate 1 μmol of reducing sugar as xylose $ml^{-1}$ $min^{-1}$ under the assay conditions.

## Purification of Recombinant Xylanase and Zymogram Analysis

The rMxyl was purified by affinity chromatography using $Ni^{2+}$-NTA agarose (Novagen, Germany) according to Verma and Satynarayana, 2012 [30]. For zymogram analysis, the recombinant protein was electrophoresed on native gel and the stripes of the gel were layered on 0.3% (w/v) xylan-agar plates of pH 9.0 at 70°C for 3 h. The gel was then removed from the plate and stained with Congo-red dye followed by destaining with 1 M NaCl for better appearance of the band showing clear zone of xylan hydrolysis.

## Biochemical Characterization of rMxyl

The kinetic characteristics of the recombinant xylanase like the effect of pH, temperature, metal ions, inhibitors and detergents on enzyme activity, thermostability and substrate specificity have been studied. The optimum pH was determined by incubating the recombinant enzyme in various buffers ranging from pH 3.0 to 12.0. Different buffers were used (pH 3.0 to 6.0, 0.1 M citrate buffer, for 7.0 and 8.0, 0.1 M sodium phosphate buffer and for 9.0–12.0, glycine-NaOH buffer). Similarly for optimum temperature, the recombinant enzyme was assayed over a range of temperatures from 40 to 100°C at pH 9.0. For thermostability, the residual xylanase activity was measured after incubating the recombinant enzyme at different temperatures (60–100) °C for 15 min to 3 h. The effects of various metal ions, inhibitors and detergents were determined by incorporating these into reaction mixtures followed by xylanase assay. The substrate specificity was quantitated by assaying the recombinant enzyme on various substrates (different sources of xylan as well as substrates other than xylan like starch, cellulose, chitin and pullulan in glycine-NaOH buffer). Kinetic properties of the recombinant enzyme ($K_m$ and $V_{max}$) on different xylans from birchwood, beechwood and oat spelt were calculated from Lineweaver-Burk double reciprocal plots.

### Saccharification of Agroresidues/Hydrolysis of Xylan

One % (w/v) standards of xylooligosaccharides (X2–X6) and agro-residues (wheat bran, corn cobs and sugarcane bagasse) were treated with recombinant xylanase (10 U – 20 U/g) to find out the hydrolysis of XOs and lignocellulosic substrates. All the substrates (wheat bran, corn cobs and sugarcane bagasse) were suspended in glycine-NaOH buffer (pH 9.0) and incubated at 80°C. Aliquots at the desired intervals were collected and analyzed on silica based TLC plates (Merck, Germany) to determine the hydrolysis products. The saccharification of agro-residues was determined using DNSA reagent [60].

### Nucleotide Sequence Accession Numbers

The nucleotide sequence of the xylanase gene is deposited at NCBI GenBank (Accession no. AFP81696).

## AUTHOR CONTRIBUTIONS

Conceived and designed the experiments: DV YK KM TS. Performed the experiments: DV YK KM TS. Analyzed the data: DV YK KM TS. Contributed reagents/materials/analysis tools: DV YK KM TS. Wrote the paper: DV YK KM TS.

## REFERENCES

1. Coughlan MP, Hazlewood GP (1993) β-1,4 D-xylan-degrading enzyme systems: biochemistry, molecular biology and applications. Biotechnol Appl Biochem 17: 259–289.
2. Hori H, Elbein AD (1985) The biosynthesis of plant cell wall polysaccharides. In: Higuchi T. (Ed. ), Biosynthesis and Biodegradation of Wood Components. Academic Press Inc. , Orlando, FL, 109–135.
3. Collins T, Gerday C, Feller G (2005) Xylanases, xylanase families and extremophilic xylanases. FEMS Microbiol Rev 29: 3–23.
4. Henrissat B (1991) A classification of glycosyl hydrolases based on amino acid sequence similarities. Biochem J 280: 309–316.
5. Paes G, Berrin JG, Beaugrand J (2012) GH11 xylanases: Structure/function/properties relationships and applications. Biotechnol Adv 30: 564–592.
6. Khandeparkar R, Bhosle NB (2007) Application of thermoalkalophilic xylanase from *Arthrobacter* sp. MTCC 5214 in biobleaching of kraft pulp. Biores Technol 98: 897–903.
7. Menon V, Prakash G, Prabhune A, Rao M (2010) Biocatalytic approach for the utilization of hemicellulose for ethanol production from agricultural residue using thermostable xylanase and thermotolerant yeast. Biores Technol 101: 5366–5373.
8. Woldesenbet F, Puneet A, Virk, Gupta N, Sharma P (2012) Effect of microwave irradiation on xylanase production from wheat bran and biobleaching of eucalyptus kraft pulp. Appl Biochem Biotechnol 167: 100–108.
9. Sunna A, Bergquist PL (2003) A gene encoding a novel extremely thermostable 1,4-β- xylanase isolated directly from an environmental DNA sample. Extremophiles 7: 63–70.
10. Sunna A, Moracci M, Rossi M, Antranikian G (1997) Glycosyl hydrolases from hyperthermophiles. Extremophiles 1: 2–13.
11. Kulkarni N, Shendye A, Rao M (1999) Molecular and biological aspects of xylanases. FEMS Microbiol Rev 23: 411–456.
12. Sharma S, Adhikari S, Satyanarayana T (2007) Alkali-thermostable and cellulose free xylanase production by an extreme thermophile

*Geobacillus thermoleovorans*. W J Microbiol Biotechnol 23: 483–490.

13. Son-Ng I, Li CW, Yeh Y, Chen PT, Chir JL, et al. (2009) A novel endoglucanase from the thermophilic bacterium *Geobacillus* sp. 70PC53 with high activity and stability over a broad range of temperatures. Extremophiles 13: 425–435.
14. Yuhong Z, Shi P, Liu W, Meng K, Bai Y, et al. (2009) Lipase diversity in glacier soil based on analysis of metagenomic DNA fragments and cell Culture. J Microbiol Biotechnol 158: 200–212.
15. Jungang L, Zhang K, Wenjun H (2010) Cloning and biochemical characterization of a novel lipolytic gene from activated sludge metagenome, and its gene product. Microbial cell factories. 9: 83.
16. Yu EY, Kwon MA, Lee M, Oh JY, Choi JE, et al. (2011) Isolation and characterization of cold-active family VIII esterases from an arctic soil metagenome. Appl Microbiol Biotechnol 90: 573–581.
17. Sunna A, Bergquist PL (2003) A gene encoding a novel extremely thermostable 1, 4-β- xylanase isolated directly from an environmental DNA sample. Extremophiles 7: 63–70.
18. Lee CC, Kibblewhite-Accinelli RE, Wagschal K, Robertson GH, Wong DWS (2006) Cloning and characterization of a cold-active xylanase enzyme from an environmental DNA library. Extremophiles 10: 295–300.
19. Hu Y, Zhang G, Li A, Chen J, Ma L (2008) Cloning and enzymatic characterization of a xylanase gene from a soil-derived metagenomic library with an efficient approach. Appl Microbiol Biotechnol 80: 823–830.
20. Yamada K, Terahara T, Kurata S, Yokomaku T, Tsuneda S, et al. (2008) Retrieval of entire genes from environmental DNA by inverse PCR with pre amplification of target genes using primers containing locked nucleic acids. Environ Microbiol 10(4): 978–987.
21. Wang G, Wang Y, Yang P, Luo H, Huang H, et al. (2010) Molecular detection and diversity of xylanase genes in alpine tundra soil. Appl Microbiol Biotechnol 87: 1383–1393.
22. Li R, Kibblewhite R, Orts WJ, Lee CC (2009) Molecular cloning and characterization of multidomain xylanase from manure library. World J Microbiol Biotechnol 25: 2071–2078.

23. Gilkes NR, Henrissat B, Kilburn DG, Miller RC Jr, Warren RA (1991) Domains in microbial 4-glycanases: sequence conservation, function, and enzyme families. Microbiol Rev 55: 303–315.

24. Brennan Y, Callen WN, Christoffersen L, Dupree P, Goubet F, et al. (2004) Unusual Microbial Xylanases from Insect Guts. Appl Environ Microbiol 70 (6): 3609–3617.

25. Shibuya H, Kaneko S, Hayashi K (2000) Enhancement of the thermostability and hydrolytic activity of xylanase by random gene shuffling. Biochem J 349: 651–656.

26. Notenboom V, Boraston AB, Williams SJ, Kilburn DG, Rose DR (2002) High-resolution crystal structures of the lectin-like xylan binding domain from *Streptomyces lividans*xylanase 10A with bound substrates reveal a novel mode of xylan binding. Biochemistry 41: 4246–4254.

27. Winterhalter C, Liebl W (1995) Two extremely thermostable xylanases of the hyperthermophilic bacterium *Thermotoga maritima* MSB8. Appl Environ Microbiol 61: 1810–1815.

28. Fontes CM, Hazlewood GP, Morag E, Hall J, Hirst BH, et al. (1995) Evidence for a general role for non-catalytic thermostabilizing domains in xylanases from thermophilic bacteria. Biochem J 307: 151–158.

29. Sakka K, Kojima Y, Kondo T, Karita S, Ohmiya K, et al. (1993) Nucleotide sequence of *Clostridium stercorarium* xynA gene encoding xylanase A: Identification of catalytic and cellulose binding domains. Biosci Biotechnol Biochem 57: 273–277.

30. Verma D, Satyanarayana T (2012) Cloning, expression and applicability of thermo-alkali-stable xylanase of *Geobacillus thermoleovorans* in generating xylooligosaccharides from agro-residues. Biores Technol 107: 333–338.

31. Guo B, Chen X, Sun C, Zhou B, Zhang Y (2009) Gene cloning, expression and characterization of a new cold-active and salt tolerant endo-β-1, 4-xylanase from marine Glaciecola mesophila KMM 241. Appl Microbiol Biotechnol 84: 1107–1115.

32. Mamo G, Delgado O, Martinez A, Mattiasson B, Kaul RJ (2006) Cloning, sequencing analysis and expression of a gene encoding

an endoxylanase from *Bacillus halodurans*S7. Mol Biotechnol 33: 149–159.

33. Shi P, Tian J, Yuan T, Liu X, Huang H, et al. (2011) *Paenibacillus* sp. Strain E18 bifunctional xylanase-glucanase with a single catalytic domain. Appl Environ Microbiol 76(11): 3620–3624.
34. Yoon HS, Han NS, Kim CH (2004) Expression of *Thermotoga maritima* endo-β-1,4 xylanase gene in *E. coli* and characterization of the recombinant enzyme. Agric Chem Biotechnol 47: 157–160.
35. Mathrani IM, Ahring BK (1992) Thermophilic and alkaliphilic xylanase from several *Dictyoglomus* isolates. Appl Microbiol Biotechnol 38: 23–27.
36. Uchino F, Fukuda O (1983) Taxonomic characteristics of an acidophilic strain of *Bacillus* producing thermophilic acidophilic amylase and thermostable xylanase. Agric Biol Chem 47: 965–967.
37. Khasin A, Alchanati I, Shoham Y (1993) Purification and characterization of a thermostable xylanase from *Bacillus stearothermophilus* T-6. Appl Environ Microbiol 59: 1725–1730.
38. Chang WS, Tsai CL, Tseng MJ (2004) Cloning and characterization of two thermostable xylanases from an alkaliphilic *Bacillus firmus*. Biochem Biophys Res Commun 319: 1017–1025.
39. Zhao Y, Luo H, Meng K, Shi P, Wang G, et al. (2011) A xylanase gene directly cloned from the genomic DNA of alkaline wastewater sludge showing application potential in the paper industry. Appl Biochem Biotechnol 165: 35–46.
40. Yoon KH (2009) Cloning of the *Bacillus subtilis* AMX-4 xylanase gene and characterization of the gene product. J Microbiol Biotechnol 19: 1514–9.
41. Maalej I, Belhaj I, Masmoudi NF, Belghith H (2009) Highly Thermostable Xylanase of the Thermophilic Fungus *Talaromyces thermophilus*: Purification and charact-erization. Appl Biochem Biotechnol. 158: 200–212.
42. Zhang GM, Huang J, Huang GR, Ma LX, Zhang XE (2007) Molecular cloning and heterologous expression of a new xylanase gene from *Plectosphaerella cucumerina*. Appl Microbiol Biotechnol 74: 339–346.

43. Liu L, Zhang G, Zhang Z, Wang S, Chen H (2011) Terminal amino acids disturb xylanase thermostability and activity. J Biol Chem. 286: 44710–44715.
44. Jalal A, Rashid N, Rasool N, Akhtar M (2009) Gene cloning and characterization of a xylanase from a newly isolated *Bacillus subtilis* strain R5. J Biosci Bioeng 107 (4): 360–365.
45. Li N, Meng K, Wang Y, Shi P, Luo H, et al. (2008) Cloning, expression, and characterization of a new xylanase with broad temperature adaptability from *Streptomyces* sp. S9. Appl Microbiol Biotechnol 80: 231–240.
46. Li R, Kibblewhite R, Orts WJ, Lee CC (2009) Molecular cloning and characterization of multidomain xylanase from manure library. World J Microbiol Biotechnol 25: 2071–2078.
47. Wang Y, Zhang H, He Y, Luo H, Yao B (2007) Characterization, gene cloning, and expression of a novel xylanase XYNB from *Streptomyces olivaceoviridis* A1. Aquaculture 267: 328–334.
48. Liu W, Shi P, Chen Q, Yang P, Wang G, et al. . (2010) Gene cloning, over expression, characterization of a xylanase from *Penicillium* sp. CGMCC 1699. Appl. Biochem. Biotechnol. 162, 1–12.
49. Matteotti C, Bauwens J, Brasseur C, Tarayre C, Thonart P, et al. (2012) Identification and characterization of a new xylanase from Gram-positive bacteria isolated from termite gut (*Reticulitermes santonensis*). Protein Expression and Purification 83: 117–127.
50. Irena R, Jacek P, Stanislaw B (2006) Isolation and properties of *Aspergillus niger* IBT-90 xylanase for bakery. Appl Microbiol Biotechnol 69: 665–671.
51. Yang Y, Zhang W, Huang J, Lin L, Lian H, et al. (2010) Purification and characterization of an extracellular xylanase from *Aspergillus niger* C3486. African J Microbiol Res 4(21): 2249–2256.
52. Vieira DS, Degreve L (2009) An insight into the thermostability of a pair of xylanases: the role of hydrogen bonds. Mol Phys 107: 59–69.
53. Kormelink FJM, Voragen AGJ (1993) Degradation of different [(glucurono)arabino] xylans by a combination of purified xylan-degrading enzymes. Appl Microbiol Biotechnol 38: 688–695.

54. Gruppen H, Hamer RJ, Voragen AGJ (1992) Water unextractable cell wall material from wheat flour. 2. Fractionation of alkali extracted polymers and comparison with water extractable arabinoxylans. J Cereal Sci 16: 53–67.

55. Mamo G, Hatti-Kaul R, Mattiasson B (2007) Fusion of carbohydrate binding modules from *Thermotoga neapolitana* with a family 10 xylanase from *Bacillus halodurans* S7. Extremophiles 11: 169–177.

56. Sun JY, LU MQ, Weng XY, Qian LC, Gu SH (2007) Expression of recombinant *Thermomonospora fusca* xylanase A in *Pichia pastoris* and xylooligosaccharides released from xylan by it. Food Chem 104: 1055–1064.

57. Vazquez MJ, Alonso JL, Dominguez H, Parajo JC (2000) Xylooligosaccharides: manufacture and applications. Trends Food Sci Technol 11: 387–393.

58. Verma D, Satynarayana T (2011) An improved protocol for DNA extraction from alkaline soil and sediment samples for constructing metagenomic libraries. Appl Biochem Biotechnol 165: 454–464.

59. Archana A, Satyanarayana T (1997) Xylanase production by thermophilic *Bacillus licheniformis* A99 in solid-state fermentation. Enzyme Microb Technol 21: 12–17.

60. Miller GL (1959) Use of dinitrosalicylic acid reagent for determination of reducing sugars. Anal Chem 31: 426–428.

# CITATIONS

## CHAPTER 6

Paolo Zucca, Antonio Rescigno, Manuela Pintus, Andrea C Rinaldi and Enrico Sanjust, Degradation of textile dyes using immobilized lignin peroxidase-like metalloporphines under mild experimental conditions, https://doi.org/10.1186/1752-153X-6-161.

## CHAPTER 7

Tang C-D, Li J-F, Wei X-H, Min R, Gao S-J, Wang J-Q, et al. (2013) Fusing a Carbohydrate-Binding Module into the *Aspergillus usamii* β-Mannanase to Improve Its Thermostability and Cellulose-Binding Capacity by *In Silico* Design. PLoS ONE 8(5): e64766. https://doi.org/10.1371/journal.pone.0064766.

## CHAPTER 8

Virk AP, Puri M, Gupta V, Capalash N, Sharma P (2013) Combined Enzymatic and Physical Deinking Methodology for Efficient Eco-Friendly Recycling of Old Newsprint. PLoS ONE 8(8): e72346. https://doi.org/10.1371/journal.pone.0072346.

## CHAPTER 9

Couturier M, Féliu J, Bozonnet S, Roussel A, Berrin J-G (2013) Molecular Engineering of Fungal GH5 and GH26 Beta-(1,4)-Mannanases toward Improvement of Enzyme Activity. PLoS ONE 8(11): e79800. https://doi.org/10. 1371/journal. pone. 0079800.

## CHAPTER 10

Yu W, Liu W, Huang H, Zheng F, Wang X, Wu Y, et al. (2014) Application of a Novel Alkali-Tolerant Thermostable DyP-Type Peroxidase from *Saccharomonospora viridis* DSM 43017 in Biobleaching of Eucalyptus Kraft Pulp. PLoS ONE 9(10): e110319. https://doi.org/10.1371/journal.pone.0110319.

## CHAPTER 11

Dhiman SS, Garg G, Sharma J, Kalia VC, Kang YC, Lee J-K (2014) Reduction in Acute Ecotoxicity of Paper Mill Effluent by Sequential Application of Xylanase and Laccase. PLoS ONE 9(7): e102581. https://doi.org/10.1371/journal.pone.0102581.

## CHAPTER 12

Verma D, Kawarabayasi Y, Miyazaki K, Satyanarayana T (2013) Cloning, Expression and Characteristics of a Novel Alkalistable and Thermostable Xylanase Encoding Gene (*Mxyl*) Retrieved from Compost-Soil Metagenome. PLoS ONE 8(1): e52459. https://doi.org/10.1371/journal.pone.0052459.

# INDEX

S



T



U



W



X



Y